CAMBRIDGE MONOGRAPHS ON PHYSICS

GENERAL EDITORS

A. HERZENBERG, PH.D.
Reader in Theoretical Physics in the University of Manchester

M. M. WOOLFSON, D.SC.
Professor of Theoretical Physics in the University of York

J. M. ZIMAN, D.PHIL., F.R.S.
Professor of Theoretical Physics in the University of Bristol

THE CHEMICAL PHYSICS OF ICE

THE CHEMICAL PHYSICS OF ICE

N. H. FLETCHER

Professor of Physics in the University of New England
Armidale, New South Wales

CAMBRIDGE
AT THE UNIVERSITY PRESS
1970

CAMBRIDGE UNIVERSITY PRESS
Cambridge, New York, Melbourne, Madrid, Cape Town, Singapore, São Paulo, Delhi

Cambridge University Press
The Edinburgh Building, Cambridge CB2 8RU, UK

Published in the United States of America by Cambridge University Press, New York

www.cambridge.org
Information on this title: www.cambridge.org/9780521112307

© Cambridge University Press 1970

This publication is in copyright. Subject to statutory exception
and to the provisions of relevant collective licensing agreements,
no reproduction of any part may take place without the written
permission of Cambridge University Press.

First published 1970
This digitally printed version 2009

A catalogue record for this publication is available from the British Library

Library of Congress Catalogue Card Number: 74–75825

ISBN 978-0-521-07597-8 hardback
ISBN 978-0-521-11230-7 paperback

CONTENTS

CHAPTER 4

Liquid water and freezing

CHAPTER 5

Crystal growth

CHAPTER 6

Thermal properties and lattice dynamics

CHAPTER 7

Point defects

CHAPTER 8

Mechanical properties

CHAPTER 9

Electrical properties

PREFACE

This book, which is about chemical physics as well as about ice, has been written with two general classes of readers in mind. The principal group, as indicated by the inclusion of the book in the present series, consists of graduate or advanced undergraduate students in physics or physical chemistry who have already taken the usual courses in quantum mechanics and solid-state physics and are looking for something on which to try their teeth. The detailed study of a reasonably simple material like ice is admirable for this purpose; not only does it employ a whole range of the techniques learnt in more formal courses but also it shows these techniques and concepts in relation to one another. For these people I have tried at all stages to show the connexions between the various properties discussed and, in particular, the way in which they all derive from the structure of the water molecule itself.

The second group is made up of those who are interested in ice for its own sake—glaciologists, cloud physicists and the like—and for these I have taken the view that what is really required is an exposition of what is now generally regarded as understood and accepted about the chemical physics of ice. I have not tried to be encyclopaedic in the fashion of Dorsey's (1940) book *Properties of Ordinary Water Substance* but have tried, rather, to produce a connected and well-documented account of what seem to me to be the major areas of interest.

I have assumed of both types of reader a reasonable familiarity with quantum mechanics and solid-state theory, though the account is one which stresses physical principles rather than theoretical models. I have, however, taken some trouble to make the account self-contained by brief reference back to the underlying physics in every case or by more detailed discussion where the application is not a standard one.

The bibliography forms an important part of the book, since everyone except the most general reader is likely to have some particular interest in the subject which he will wish to pursue further. It is not an exhaustive bibliography but I have tried to include those papers which will be found most useful for further study and hope that no really significant citation has been omitted. A casual glance will shown that most of the references are quite recent, and there are two reasons for this. In the first place, much of our detailed understanding of the structure and properties of ice has been attained during the past twenty years so that earlier work, while reporting useful results, does so against a rather different background, and many of the measurements have in any case been repeated more carefully since; secondly, Dorsey's compilation gives a nearly complete list of some 2000 references prior to 1938, including 600 specifically on ice, and it would be wasteful to repeat any but the most important here.

A book such as this naturally draws its subject-matter from the combined endeavours of a very large number of individual workers and I should like to thank those who have given me generous permission to reproduce drawings or other material from their publications and particularly those who have informed me of new results prior to publication. My best thanks are also due to my research students, whose critical discussions have been a constant stimulus, to the Australian Research Grants Committee which supports our work on statistical solids and to Miss Pat Bannister for her excellent work in preparing the typescript.

N. H. FLETCHER

Armidale, December 1968

NOTE ON UNITS

I have tried at every stage to use the units which seem to me to be most appropriate for the quantity described. This is, however, largely a question of personal background and preference and I cannot hope to have pleased everyone. The table and nomogram given below may be useful for those who like to think in terms of other units.

	eV	kcal mole^{-1}	erg	cm^{-1}	degK
1 eV	1	23·0	$1{\cdot}6 \times 10^{-12}$	$8{\cdot}05 \times 10^{3}$	$1{\cdot}16 \times 10^{4}$
1 kcal mole^{-1}	$4{\cdot}35 \times 10^{-2}$	1	$6{\cdot}96 \times 10^{-14}$	$3{\cdot}50 \times 10^{2}$	$5{\cdot}04 \times 10^{2}$
1 erg	$6{\cdot}25 \times 10^{11}$	$1{\cdot}44 \times 10^{13}$	1	$5{\cdot}03 \times 10^{15}$	$7{\cdot}25 \times 10^{15}$
1 cm^{-1}	$1{\cdot}24 \times 10^{-4}$	$2{\cdot}86 \times 10^{-3}$	$1{\cdot}99 \times 10^{-16}$	1	1·44
1 degK	$8{\cdot}6 \times 10^{-5}$	$1{\cdot}98 \times 10^{-3}$	$1{\cdot}38 \times 10^{-16}$	$6{\cdot}94 \times 10^{-1}$	1

CHAPTER 1

THE WATER MOLECULE

To gain a proper understanding of the behaviour of a complex system we must first appreciate the structure and properties of the elementary units of which it is composed. In the study of ice this means that we must begin with a study of the water molecule, for it is from the individuality of the structure of that molecule that most of the unusual properties of ice and water arise. Without such a relation back to the fundamentals of molecular structure, the study of a particular material becomes simply a catalogue of its properties—useful, no doubt, but not very illuminating. In this book we shall try, at all stages, to show this relation so that a coherent picture emerges. Similar pictures can be built up for all solids; the outlines, it is true, have many variations but they all follow in the same sort of way from the basic elements of which they are built.

Water is among the simplest of molecules and for that reason if no other it has been the subject of a large amount of theoretical work, computation and experiment. Molecular structure is not a simple subject, however, and our understanding is still far from complete, but it is enough to give a picture of the water molecule which is reasonably accurate and sufficient for our present purposes. There are many aspects of its properties which we shall not need to explore, since the molecule itself is not our primary interest, and for these the reader is referred to the standard works by Coulson (1961), Herzberg (1945, 1966), Pauling (1960) and Slater (1963).

1.1. Electronic structure

The water molecule is triangular in shape. This much we know from specific heat data on water vapour near room temperature. The experimental value of C_p at 100 °C is 8·65 cal mole^{-1} deg^{-1}, which is equivalent to a value for C_v of about $3{\cdot}3k$ per molecule. Since the vibrational frequencies are too high to be greatly excited at this temperature, we must interpret this specific heat as a contribution $\frac{3}{2}k$ from translation, together with $\frac{1}{2}k$ from rotations about each of three axes for which the molecule has an appreciable moment of inertia. This rules out a linear molecule, which is the only alternative to one of triangular shape. The same conclusion is reached in more detail, as we shall see later, from spectroscopic studies.

Our problem now is to see how this shape arises from the electronic structure of the atoms involved and to find out what we can about wave functions, energies, charge distributions and so on, all of which will be needed for an understanding of the interactions of these molecules in the liquid and solid states. Even using the Born–Oppenheimer approximation, which allows us first to solve the electronic problem with the nuclei fixed and then to use this result to determine the effective potential in which the nuclei move, exact solution of the Schrödinger equation is out of the question. It is possible, however, with relatively little labour, to see how the particular structure of the water molecule comes about and then, by refining this crude model, to calculate relevant quantities quite accurately.

The oxygen atom has the electronic structure $(1s)^2(2s)^2(2p)^4$ so that we may expect, to a first approximation, that only electrons in the incomplete $2p$ sub-shell will be involved in chemical binding. More specifically, if we consider $(2p)^4$ to be $(2p_x)(2p_y)(2p_z)^2$, then only the two electrons in the $2p_x$ and $2p_y$ orbitals will be able to form bonds, while the two in the $2p_z$ orbital form a 'lone pair'. To form a molecule H_2O in its equilibrium state we place the two hydrogen atoms on the x-and y-axes respectively so that their wave functions overlap as strongly as possible with the oxygen $2p_x$ and $2p_y$, for in this way the energy can be minimized. The new bonding orbitals created by this overlap have the general form

$$\left.\begin{aligned}\psi_{\mathrm{I}} &= \mathrm{O}(2p_x)+\eta\mathrm{H}_{\mathrm{I}}(1s),\\ \psi_{\mathrm{II}} &= \mathrm{O}(2p_y)+\eta\mathrm{H}_{\mathrm{II}}(1s),\end{aligned}\right\} \qquad (1.1)$$

and are shown schematically in fig. 1.1(*a*). Each of the orbitals contains two electrons with opposite spins. The parameter η is positive, leading to a concentration of charge in the low-potential region between the O and H nuclei and hence to a finite binding energy for the molecule. η also measures the polarity of the bonds in the sense that, if η is small, most of the electronic charge on the hydrogen atom has been transferred to the oxygen atom.

The equilibrium O–H distances are determined by a compromise between the extra electronic overlap, with its consequent lowering of energy, which can be achieved by close approach and the increase in energy due to Coulomb repulsion between the nuclei. Since there must be some electronic overlap for bonding we might expect an O–H distance comparable to an atomic 'radius' which is of the order of 1 Å.

The configuration predicted by this very simple approach is triangular, with an H–O–H angle of 90°. However, it can easily be seen that, because the two O–H bonds are not electrically neutral, they will tend to repel each other so that an equilibrium angle greater than 90° is to be expected. This leads to an adjustment of the electronic configuration to keep the bonding overlap as great as possible, this being accomplished by the admixture of some O($2s$), O($2p_x$) and O($2p_y$) contribution into the bonding orbitals given by (1.1). This extra hybridization of the oxygen orbitals reacts in turn upon the lone-pair O($2p_z$) orbitals and on the O($2s$) orbitals so that they too become hybrids. The final equilibrium configuration is thus roughly tetrahedral, with the oxygen nucleus at the centre, protons at two vertices and an electron distribution with the lone-pair orbitals directed towards the two remaining vertices as shown in fig. 1.1 (*b*).

Modern calculations of the structure of the water molecule amount to a refinement of the approach outlined above. The usual practice is to assume a particular form for the electronic wave functions and then to vary this until the energy is a minimum. This either can be done by a 'brute-force' method, using a large set of trial functions of relatively simple form but each containing several adjustable parameters, or can start from a limited set of more complicated multi-centre functions closely related to the symmetry of the water molecule.

The chief attraction of the first approach, as illustrated by the work of Bishop *et al.* (1963, 1966) and of Moccia (1964), is that it leads to a very good value for the total molecular energy, and hence for certain other properties of the molecule, without involving complicated multi-centre integrals. Its disadvantage is that it

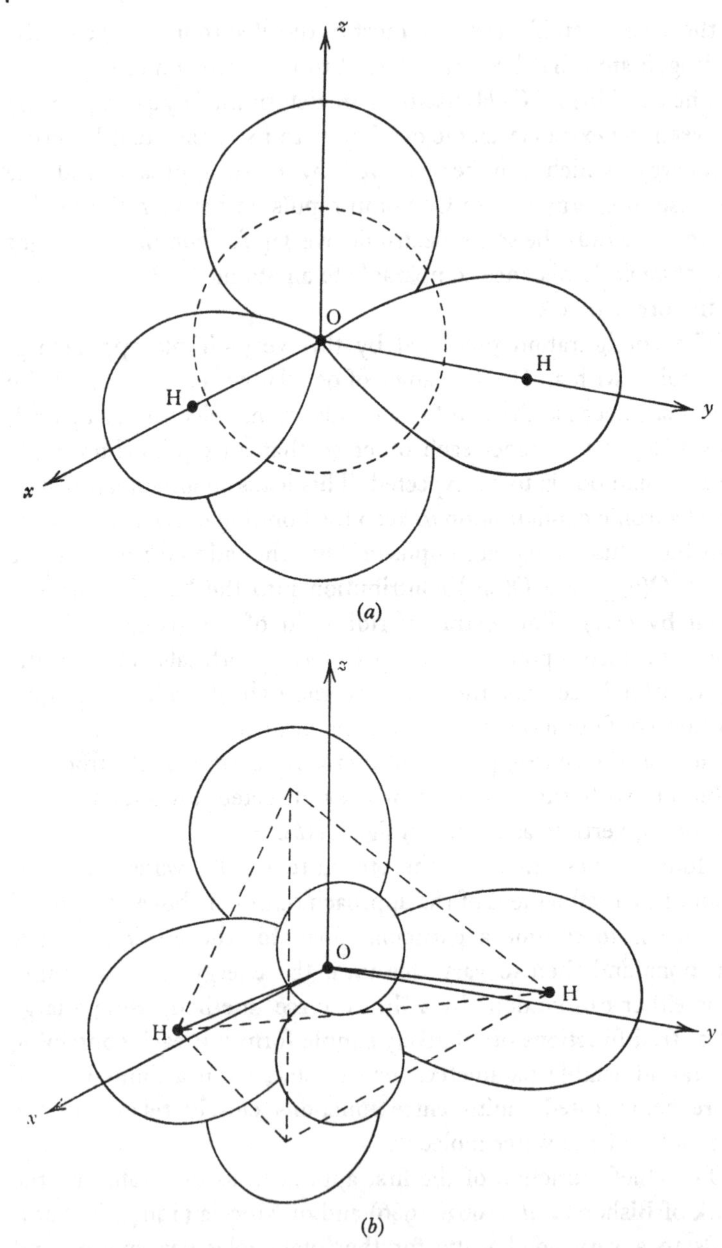

Fig. 1.1. Schematic representation of orbital configurations in the water molecule, (*a*) according to the simplified scheme of (1.1) in which O(2*s*) and O($2p_z$) electrons do not participate in bond formation, (*b*) with approximately tetrahedral hybridization. The actual calculated electron density is shown in fig. 1.3.

does not directly yield information on the electronic energy levels of the molecule and that its results cannot be simply related to those of simple pictures like that discussed above. A calculation based upon wave functions whose symmetry is related to that of the molecule itself has the advantage that it gives explicit electronic energy levels and is readily interpreted in terms of the structure of the molecule, but pays a heavy penalty because of the difficulties involved in evaluating the multi-centre integrals involved.

The symmetry of the molecule, as shown in fig. 1.2, is defined by the group symbol C_{2v}, which means that it has a twofold axis of rotation symmetry (the line bisecting the H–O–H angle), C_2, together with a plane of reflexion symmetry, v, passing through this axis and normal to the plane of the molecule. These two symmetry elements imply, in addition, a plane of reflexion symmetry, v', passing through the three nuclei as can easily be seen from the figure. For this simple symmetry the molecular wave functions can then be classified according to whether or not they change sign under these reflexion operations. Thus functions of type A_1 are invariant under both reflexions v and v', type A_2 change sign under both, type B_1 change sign only under v' and B_2 only under v.

If we make the approximate assumption that molecular orbitals can be formed from linear combinations of atomic orbitals, as was done in (1.1), then, choosing axes so that the z-axis bisects the H–O–H angle and the y-axis is in the plane of the molecule as shown in fig. 1.2, the possible orbitals are

$$\left.\begin{aligned} \psi(A_1) &= a_1\mathrm{O}(1s)+a_2\mathrm{O}(2s)+a_3\mathrm{O}(2p_z)+a_4[\mathrm{H_I}(1s)+\mathrm{H_{II}}(1s)], \\ \psi(B_1) &= \mathrm{O}(2p_x), \\ \psi(B_2) &= b_1\mathrm{O}(2p_y)+b_2[\mathrm{H_I}(1s)-\mathrm{H_{II}}(1s)]. \end{aligned}\right\} \tag{1.2}$$

There are no molecular orbitals of type A_2 unless we include higher atomic orbitals. There are actually four different orbitals $\psi(A_1)$ corresponding to four independent choices of the coefficients a_i, and two different orbitals $\psi(B_2)$.

The structure of the molecule has been studied on this basis by Pople (1950), by Ellison & Shull (1953, 1955) and by McWeeny & Ohno (1960), using slightly different approximations. The atomic orbitals in particular were not included exactly but approximated by Slater functions which, for principal quantum number n and angular momentum $(\mathbf{l}, m)$, have the form

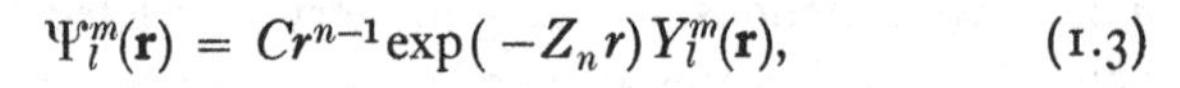

$$\Psi_l^m(\mathbf{r}) = Cr^{n-1}\exp(-Z_n r)\,Y_l^m(\mathbf{r}), \qquad (1.3)$$

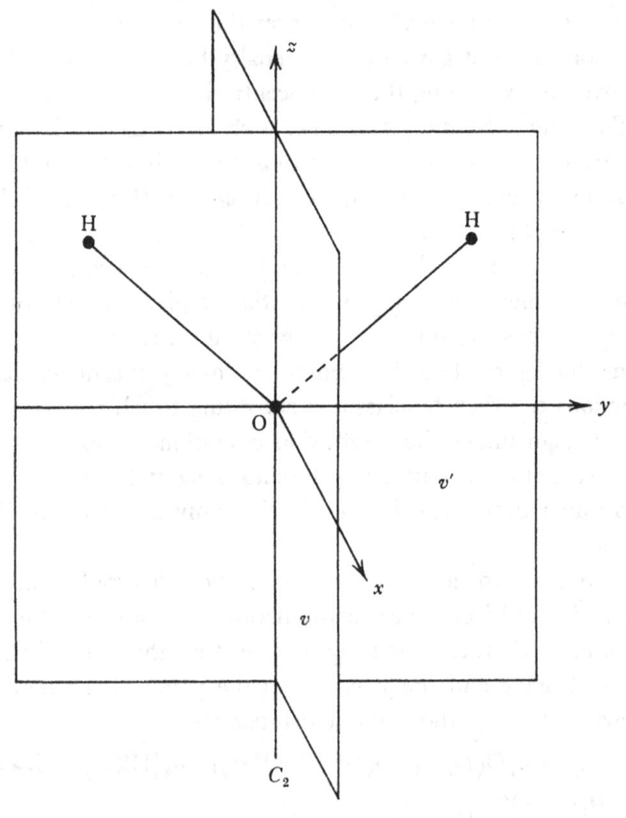

Fig. 1.2. Axes and symmetry elements for the water molecule.

where Y_l^m is a spherical harmonic and Z_n is an effective nuclear charge divided by the principal quantum number n, the Z_n being determined from study of atomic wave functions. In most of these calculations the configuration of the nuclei is assumed to be that determined from experiment, partly to reduce the amount of calculation involved and partly because it is rather difficult to reproduce the experimental bond angle (105°) by calculation.

The procedure involved in calculating the electronic structure is essentially to minimize the energy with respect to the parameters a_i and b_i appearing in the wave functions (1.2). This leads to the conclusion that the ground-state configuration is

$$(A_1^{(1)})^2(A_1^{(2)})^2(B_2^{(1)})^2(A_1^{(3)})^2(B_1^{(1)})^2, \tag{1.4}$$

where the orbitals have been written in ascending order of energy. The orbital $A_1^{(1)}$ is essentially the O(1s) state, $A_1^{(2)}$ is a bonding combination of $O(2p_z)$ with $H_I(1s)+H_{II}(1s)$ and $B_2^{(1)}$ a bonding combination of $O(2p_y)$ with $H_I(1s)-H_{II}(1s)$. The orbital $A_1^{(3)}$ is a lone-pair orbital made up from a combination of O(2s) with $O(2p_z)$ and directed along the z-axis away from the hydrogens, while $B_1^{(1)}$ is another lone-pair orbital $O(2p_x)$ directed out of the plane of the molecule. The remaining possible orbitals $A_1^{(4)}$ and $B_2^{(2)}$ are antibonding and are unoccupied in the ground state.

TABLE 1.1. *Approximate orbital energies for the water molecule (Ellison & Shull, 1955)*

	Energies (eV)	
	Calculated	Experimental
$A_1^{(1)}$	−557·3	—
$A_1^{(2)}$	−37·19	—
$B_2^{(2)}$	−18·55	−16·2 ± 0·3
$A_1^{(3)}$	−13·20	−14·5 ± 0·3
$B_1^{(1)}$	−11·79	−12·6 ± 0·1
$A_1^{(4)}$	+13·7	—
$B_2^{(2)}$	+15·9	—

The approximate energies of these orbitals, as found by Ellison and Shull, are given in table 1.1, together with experimental values found by electron impact ionization potentials and from spectroscopic data. It is clear that the agreement, while not perfect, is quite good.

The total energy of the molecule cannot be obtained simply by summing the energies of the separate orbitals, because of the cross-terms between them. The total electronic energy calculated by Ellison and Shull was −2312·4 eV. Adding the nuclear repulsion energy of 249·91 eV, this gives a total energy for the molecule of −2062·5 eV, which is within 1 per cent of the experimental value,

−2080·6 eV. The binding energy of the molecule, obtained by comparing the molecular energy with that calculated for the separated atoms using the same basis functions, is not in such good agreement with experiment, the calculated 7·7 eV being considerably less than the experimental value of 10·06 eV.

More recent calculations, as summarized by Bishop & Randić (1966), have improved these energy values slightly, though the energy of the best calculation is still about 0·5 per cent above the experimental value. There is little to choose in this respect between calculations using symmetry orbitals and those based upon single-centre functions.

The calculated wave functions, and hence the electron density distribution, may differ slightly from their true form because the variational method gives an energy value which is not critically dependent upon the exact form of the trial wave functions used in its evaluation. It is satisfying to find, however, that the electronic charge distribution, as measured by the various multipole moments of the molecule, does not vary greatly from the wave functions of one calculation to those of another. If the oxygen nucleus is chosen as origin with the z-axis bisecting the H–O–H angle, then the dipole moment is defined by

$$\boldsymbol{\mu} = \int \mathbf{r}\rho(\mathbf{r})\,d\mathbf{r} \tag{1.5}$$

and the quadrupole and octupole moments by

$$\left.\begin{aligned} \mathbf{Q}_{\alpha\beta} &= \int r_\alpha r_\beta \rho(\mathbf{r})\,d\mathbf{r}, \\ \mathbf{O}_{\alpha\beta\gamma} &= \int r_\alpha r_\beta r_\gamma \rho(\mathbf{r})\,d\mathbf{r}, \end{aligned}\right\} \tag{1.6}$$

respectively, where $\rho(\mathbf{r})$ is the electron density and the subscripts specify cartesian components. The ranges of values calculated by Glaeser & Coulson (1965) from the wave functions of different authors are given in table 1.2. The experimental dipole moment is $\mu = 1{\cdot}84 \times 10^{-18}$ e.s.u., to which the calculated values give a reasonably good approximation. The higher moments are not known from experiment but the consistency of the calculations suggests that the values given in table 1.2 are probably correct in sign and in general magnitude. It should be remarked in passing that some authors use definitions of higher moments rather different from those given in (1.6).

TABLE 1.2. *Calculated multipole moments for the water molecule, assuming a bond angle of* 105° (*Glaeser & Coulson*, 1965)

Dipole
$\mu = 1{\cdot}42$ to $1{\cdot}72 \times 10^{-18}$ e.s.u.
Quadrupole
$Q_{xx} = -6{\cdot}51$ to $-6{\cdot}56 \times 10^{-26}$ e.s.u.
$Q_{yy} = -5{\cdot}03$ to $-5{\cdot}48 \times 10^{-26}$ e.s.u.
$Q_{zz} = -5{\cdot}42$ to $-5{\cdot}80 \times 10^{-26}$ e.s.u.
Octupole
$O_{xxz} = -1{\cdot}04$ to $-1{\cdot}30 \times 10^{-34}$ e.s.u.
$O_{yyz} = -0{\cdot}38$ to $-0{\cdot}71 \times 10^{-34}$ e.s.u.
$O_{zzz} = -2{\cdot}57$ to $-2{\cdot}77 \times 10^{-34}$ e.s.u.

It is interesting to see how these electrical moments, and particularly the dipole moment, can be thought of in terms of contributions from different parts of the molecule. The nuclei are, of course, easily treated but it is helpful to think of the electronic contribution, not in terms of symmetry orbitals like (1.2) but rather as localized orbitals which can be pictured as O–H bonds and lone pairs. This can be done by performing a unitary transformation upon the wave function specified in terms of components (1.2) to express it in terms of new linear combinations of these functions which have the required localization and direction. The details of the transformation have been discussed by Pople (1950) and by McWeeny & Ohno (1960). The resulting bond orbitals look rather like extensions of the simple forms given in (1.1) and are quite similar to the tetrahedral sp^3 hybrids used in discussing the bonds formed by carbon atoms, except that the hybridization is rather less developed.

The result of analysis in these terms is to see that a surprisingly large fraction of the total dipole moment is contributed by the lone-pair electrons of the oxygen atom (Coulson, 1951; Pople, 1953; Duncan & Pople, 1953). These calculations are on a somewhat different basis from the energy-minimizing calculations we have been discussing, the procedure being rather to determine the parameters in a set of trial wave functions by requiring that they reproduce the experimental value for the dipole moment. Despite disagreement in detail, it seems that the electron distribution in the O–H bonds balances, or even outbalances, the dipole moment due to the nuclei, so that the magnitude of the total molecular moment is largely determined by the lone pairs.

The values given in table 1.2 may be useful in calculating the interactions of water molecules in the vapour phase when they are not close together. They cannot, however, properly be used to calculate interactions in the liquid or solid because they will be greatly upset by induced moments due to neighbouring molecules. We shall return to this later. It is worth while, however, to compare the results of table 1.2 with the multipole moments deduced from simplified point-charge models for the molecule. The most natural of these models treats the molecule as a tetrahedral arrangement of two positive point charges, representing the protons, and two equal negative point charges representing the lone pairs (Rowlinson, 1951; Bjerrum, 1951). Adjustment of charge and tetrahedron size gives the correct dipole moment but the quadrupole moments are quite wrong, $\mathbf{Q}_{xx}$ being an order of magnitude too small and $\mathbf{Q}_{yy}$ and $\mathbf{Q}_{zz}$ of the wrong sign (Glaeser & Coulson, 1965). Similar objections apply to the triangular point-charge models of Bernal & Fowler (1933) and of Rowlinson (1951) and it seems that such models can be justified only for the very crudest calculations.

The distribution of electronic charge density in the water molecule is shown in fig. 1.3, which was calculated by Bader on the basis of a method described by Bader & Jones (1963) in which a limited set of basis functions is used and the parameters involved are varied so that the net force on each nucleus is made zero. The electron density distribution determined in this way is very similar to that found from calculations like those of Ellison and Shull but varies somewhat from the results found by Pople, the tetrahedral hybridization being less developed. The electron density at the proton site is clearly quite large, actually about 20 per cent greater than in an isolated hydrogen atom, and the total density contours are quite distorted in this region. In contrast, the lone pairs are not very localized at all and show up rather as a diffuse concentration of negative charge in a ridge perpendicular to the plane of the molecule.

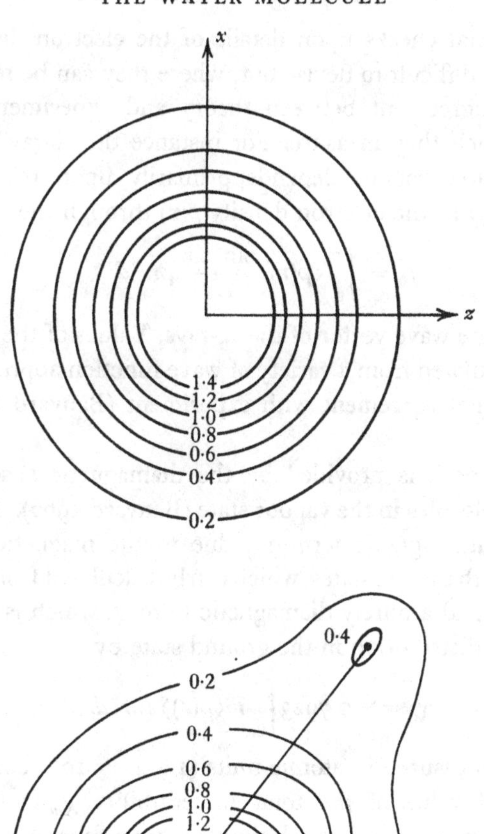

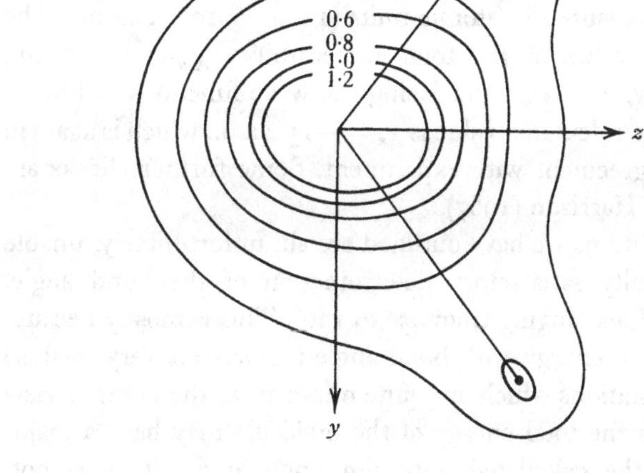

Fig. 1.3. Electron density contours in the water molecule, as calculated by Bader. Densities are given in atomic units (1 a.u. = ea_0^{-3} = $6{\cdot}749\,e\text{Å}^{-3}$).

Experimental checks upon details of the electron density distribution are difficult to devise but, where they can be made, they show good agreement between theory and experiment for the property which they measure. For instance the X-ray scattering factor of water vapour depends primarily upon the spherical average $\langle\rho(r)\rangle$ of the electron density $\rho(\mathbf{r})$ through the relation

$$f_0 = \int_0^\infty \langle\rho(r)\rangle \frac{\sin Kr}{Kr} 4\pi r^2 \, dr, \quad (1.7)$$

where **K** is the wave vector of the X-rays. Values of the scattering factor f_0 calculated from a variety of wave function approximations are in excellent agreement with experiment (Banyard & March, 1957).

Another check is provided by the diamagnetic susceptibility χ_m of the molecules in the vapour state (Banyard 1960). This is the sum of a paramagnetic term χ_{hf} due to the magnetic moment associated with excited states, which can be calculated from spectroscopic data, and a purely diamagnetic term χ_r which is related to the electron distribution in the ground state by

$$\chi_r = -0{\cdot}7923 \int_0^\infty r^2 \langle\rho(r)\rangle \, 4\pi r^2 \, dr, \quad (1.8)$$

where χ_r is measured in atomic units (1 a.u. $= 10^{-6}$ e.m.u.). The experimental value of the total susceptibility χ_m is $-13{\cdot}0$ a.u., which gives $\chi_r = -14{\cdot}46$ a.u. Using the wave functions of Ellison and Shull, the calculated value is $\chi_r = -15{\cdot}2$ a.u., which is again in quite good agreement with experiment. Some further checks are described by Harrison (1967).

The calculations we have outlined are all, unfortunately, unable to give a really satisfactory determination of the bond angle, calculated values ranging from 95° to 120°. This is mostly because the variation of energy with bond angle is relatively very small so that approximations which are quite minor from the point of view of calculating the total energy of the molecule may have a major effect upon the calculated optimum bond angle. This is not, however, an uncertainty in principle about the factors determining the angle but rather a reflexion of the difficulties involved in making precise calculations from first principles in a system which is far from simple.

1.2. Molecular vibrations

The electronic calculations we have discussed determine the electron configuration with the three nuclei held fixed. If the calculations have been carried through accurately using the observed O–H distances and H–O–H bond angle, or if these two quantities have been adjusted to minimize the total energy, then the calculated state will be an equilibrium one. Any relative displacement of the nuclei will be followed by the electrons, in the sense that they adjust almost immediately to their new configuration of lowest energy, but the total energy of the molecule will be increased. The nuclei can therefore oscillate in various ways about their equilibrium positions.

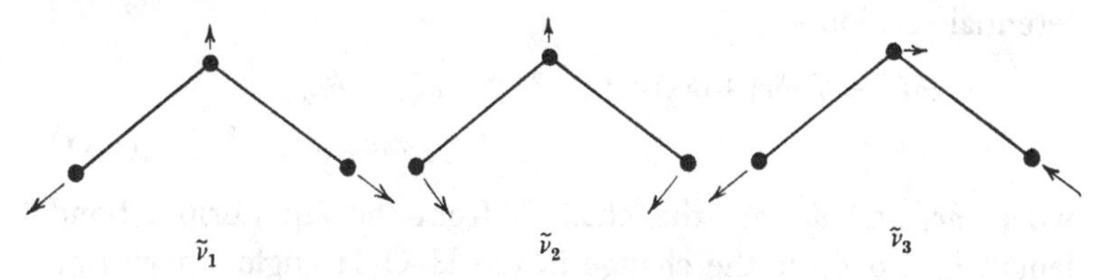

Fig. 1.4. The three normal modes for vibration of the water molecule.

As with the electronic structure, it is most convenient to consider these oscillations in terms of normal modes with symmetry related to that of the molecule. Since there are three nuclei involved, the molecule has $3 \times 3 = 9$ degrees of freedom, of which three can be taken to describe the position of the centre of mass and three the orientation of the molecule. This leaves three coordinates to describe the configuration of the molecule, which implies three possible normal modes. These three modes, which have the symmetry properties outlined for the group C_{2v} in the last section, are illustrated in fig. 1.4. The first two are symmetric under all operations of the point group, while the third is antisymmetric under the rotation C_2 or the reflexion v. The mode $\tilde{\nu}_1$ can be thought of as a symmetrical combination (in-phase) of two O–H stretching vibrations while $\tilde{\nu}_3$ is an out-of-phase combination of the same vibrations. The mode $\tilde{\nu}_2$ represents a distortion of the H–O–H angle. In all cases, though the bulk of the oscillation can be attributed to the protons, the oxygen nucleus must move slightly to maintain the position of the centre of mass. The frequencies (in wave numbers) associated with these three normal modes are

$$\tilde{\nu}_1 = 3652, \quad \tilde{\nu}_2 = 1595, \quad \tilde{\nu}_3 = 3756 \text{ cm}^{-1}, \qquad (1.9)$$

while the wave numbers derived for the associated harmonic vibrations of infinitesimal amplitude are

$$\tilde{\nu}'_1 = 3832, \quad \tilde{\nu}'_2 = 1648, \quad \tilde{\nu}'_3 = 3942 \text{ cm}^{-1}. \qquad (1.10)$$

It is possible, in principle, to derive the force constants for these oscillations from the electronic calculations of the previous section. We can write the energy U of the molecule in terms of symmetry co-ordinates or as a general quadratic function of the relative co-ordinates of the three nuclei. In this latter form we have the differential relation

$$2\delta U = f_r(\delta r_1^2 + \delta r_2^2) + f_\phi(r_0\,\delta\phi)^2 + 2f_{rr}\,\delta r_1\,\delta r_2 + 2f_{r\phi}r_0\,\delta\phi\,(\delta r_1 + \delta r_2), \qquad (1.11)$$

where δr_1 and δr_2 are the changes from the equilibrium bond length r_0 and $\delta\phi$ is the change in the H–O–H angle. In writing (1.11) the contribution of cubic and higher terms has been neglected. The force constants f can then be found by direct energy calculations with particular distortions of the bond angle. The force constants can also be derived from infrared spectral data, possible ambiguities being resolved by comparison with spectra of HDO and D_2O, which should have the same potential energy function but different frequencies because of the different nuclear masses. The force constants found by Mills (1963) in this way are

$$\left.\begin{aligned} f_r &= 8{\cdot}45 \times 10^5 \text{ dyne cm}^{-1}, \\ f_{rr} &= -0{\cdot}09 \times 10^5 \text{ dyne cm}^{-1}, \\ f_\phi &= 0{\cdot}76 \times 10^5 \text{ dyne cm}^{-1}, \\ f_{r\phi} &= 0{\cdot}26 \times 10^5 \text{ dyne cm}^{-1}. \end{aligned}\right\} \qquad (1.12)$$

The value of $2(f_r + f_{rr})$ calculated from first principles by Bishop and Randić is in very good agreement.

It is interesting to note that the coefficient f_ϕ, corresponding to distortion of the H–O–H angle, is an order of magnitude smaller than the bond-stretching coefficient f_r. This explains, in part, the difficulty experienced in determining the bond angle by minimizing the total energy. From (1.11) and (1.12), a change of 10° in the H–O–H angle changes the total energy of the molecule by only 0·1 per cent, which is well below the limit of reliability of the variational calculations. The difference between f_r and f_ϕ is also reflected in the relative frequencies of the stretching and bending modes, $\tilde{\nu}_1$ and $\tilde{\nu}_2$. Coupling between stretching of the individual O–H bonds is small, as is shown by the magnitude of f_{rr}.

1.3. Spectral properties

To conclude this survey of those properties of the water molecule which will be important in our later discussion, we give a brief review of the transitions which can occur between the different energy states of the molecule and the spectral absorptions which are associated with them.

The greatest energy changes are, of course, those associated with alteration of the electronic state of the molecule. Only transitions from the ground state will concern us here (though of course transitions between excited states can occur, for example, in electrical discharges). Since the molecule is chemically stable, the first electronic transition occurs at a wavelength well into the ultraviolet region and leads to a broad continuum of rotational and vibrational levels extending to shorter wavelengths. The absorption curve of water vapour in the region from 1150 to 1900 Å, as determined by Schmitt & Brehm (1966) is shown in fig. 1.5, similar measurements being obtained by Watanabe & Zelikoff (1953). We shall not discuss this curve in detail here.

In the visible region of the spectrum water vapour is transparent and all further absorptions of interest occur in the infrared or at even longer wavelengths. These are associated with transitions between vibrational levels of the molecule, the fundamental modes for which are shown in fig. 1.4, and have a fine structure dependent upon the rotational levels involved. Since each of the three normal modes has a direct effect upon the dipole moment of the molecule, they all lead to absorption bands. Because the interatomic potentials have appreciable anharmonic components from terms of cubic or higher order in the displacements, the relation between successive energy levels is not simple and the observed absorptions are not centred exactly on the zero-order frequencies given in (1.9) (Herzberg, 1945, pp. 201–9).

The bands which are of particular interest are those which lie in the near-infrared region and which can be fairly easily related to particular vibrational transitions, for the changes in these bands when the molecules are closely interacting in a liquid or solid phase

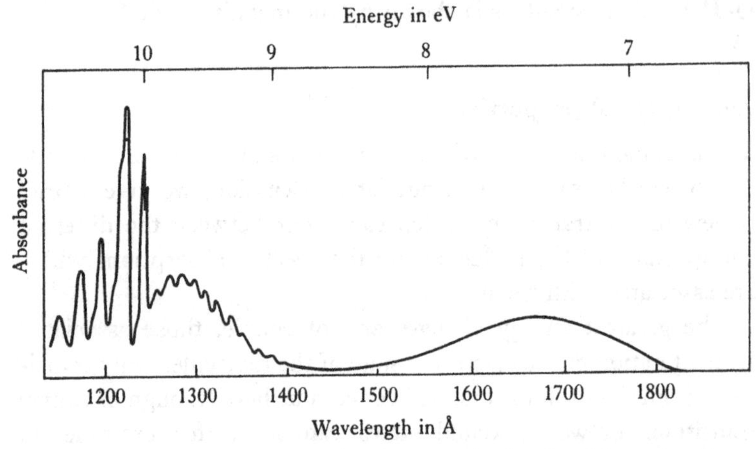

Fig. 1.5. Absorption spectrum of water vapour in the ultraviolet region (from Schmitt & Brehm, 1966: *Applied Optics* **5**, 1111–16, fig. 7; © The Optical Society of America).

can give useful information about intermolecular forces. The general form of these near-infrared bands is shown in fig. 1.6, the bond-bending mode $\tilde{\nu}_2$ contributing the absorption near 6 μ while the two stretching modes $\tilde{\nu}_1$ and $\tilde{\nu}_3$ overlap in the strongly absorbing band just below 3 μ. The other absorptions which can be seen are due to transitions to higher vibrational states representing combinations of the fundamental modes.

Within each absorption band the absorption is not continuous but has the form of a series of closely spaced lines representing transitions between different rotational levels of the molecule. Because the molecule H_2O is an asymmetric top, in the classical sense, meaning that its three principal moments of inertia are all unequal, the line spacing is very irregular. Several bands have, however, been analysed (Herzberg, 1945, pp. 484–9) and from them the principal moments of inertia and hence the geometry of the molecule deduced. In the equilibrium configuration the principal moments have values 1·024, 1·921 and $2{\cdot}947 \times 10^{-40}$ g cm^2 respectively, giving for the equilibrium O–H distance and (H–O–H) angle

$$(\text{O–H})_e = 0{\cdot}958 \times 10^{-8}\ \text{cm}; \quad (\text{H–O–H})_e = 104^\circ\ 27'. \quad (1.13)$$

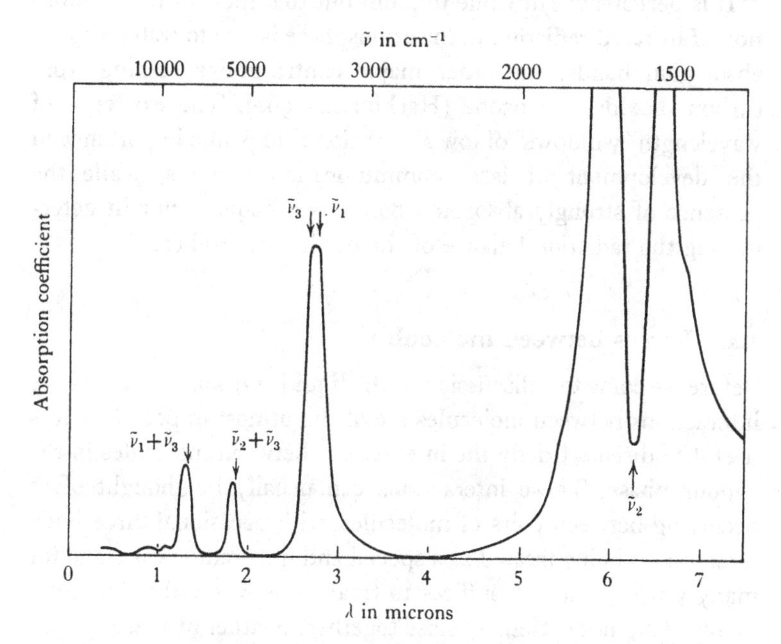

Fig. 1.6. General form of the absorption spectrum of water vapour in the near-infrared region (from data given by Hackforth, 1960).

In the lowest vibrational state of the molecule the values are slightly different:

$$(\text{O–H})_0 = 0{\cdot}957 \times 10^{-8}\ \text{cm}; \quad (\text{H–O–H})_0 = 105^\circ\ 3'. \quad (1.14)$$

In the far-infrared spectral region water vapour can absorb radiation by transition between different rotational levels without any vibrational or electronic changes. Such a pure rotational spectrum can only be exhibited by a molecule with a permanent dipole moment. The rotational spectrum has been measured for wavelengths up to 2 mm (Herzberg, 1945, p. 58; Furashov, 1966), where this type of absorption ceases and, in good agreement with theory, consists of relatively sharp lines distributed in what looks like random fashion. If the absorption coefficient of the vapour is measured, in the presence of air at atmospheric pressure and with a relatively large bandwidth, then its magnitude is of order 10^4 per cm of equivalent liquid water path length, which is a very high value.

It is perhaps worth while to point out that most of the attenuation of infrared radiation in the atmosphere is due to water-vapour absorption bands, the other major contributions coming from carbon dioxide and ozone (Hackforth, 1960). The existence of wavelength 'windows' of low absorption is of prime importance in the development of laser communication systems, while the presence of strongly absorbing bands is a major factor in determining the radiation balance of the earth's atmosphere.

1.4. Forces between molecules

Before we come to a discussion of the liquid and solid states, where interactions between molecules are of the utmost importance, it is useful to discuss briefly the interactions between molecules in the vapour phase. These interactions can usually be thought of as occurring between pairs of molecules, with occasional three-body interactions being treated as a special and uncommon case, and for many purposes it also suffices to treat cases where the molecules involved are not extremely close together. Neither of these simplifications can be made for molecules in a condensed phase.

Whilst the interaction forces between two molecules cannot rigorously be split up into contributions arising from different effects, it is often a help to make some such separation and we shall do it here. If we deal only with simple, closed-shell diamagnetic molecules, of which the water molecule is one, then the interaction energy between two such molecules can be considered as arising from four causes: (i) electrostatic interaction between the permanent multipole moments of the two molecules; (ii) electrostatic interaction between the permanent multipole moments of one molecule and the moments which these induce in the other molecule because of its finite polarizability; (iii) an interaction of quantum origin between the instantaneous fluctuating dipole moments of the two molecules (dispersion forces); (iv) an interaction of quantum origin, due to the exclusion principle for electrons, which only becomes important when the electron clouds of the two molecules begin to overlap (exchange forces).

All these interactions are taken into account automatically in a proper quantum-mechanical treatment (Buckingham, 1965) but it is instructive to examine them separately.

The interaction between permanent multipole moments is quite straightforward to calculate. Using the definitions of the multipole moments given in (1.6) and choosing symmetry axes as in fig. 1.2, the electrostatic potential due to an isolated molecule can be written (Coulson & Eisenberg, 1966*a*) as

$$V(r, \theta, \phi) = \frac{\mu \cos\theta}{r^2} + \frac{1}{2r^3}[(\mathbf{Q}_{zz} - \mathbf{Q}_{xx})(1 - 3\sin^2\theta\cos^2\phi) + (\mathbf{Q}_{zz} - \mathbf{Q}_{yy})(1 - 3\sin^2\theta\sin^2\phi)] + \dots \quad (1.15)$$

Since the molecule is uncharged there is no r^{-1} term, the dipole term falls off as r^{-2} and the quadrupole terms as r^{-3}. Another similar molecule at the point (r, θ, ϕ) and with an orientation specified by new symmetry axes x', y', z' will have an energy determined by the interaction of its dipole moment μ with the field component $-\partial V/\partial z'$ and of its quadrupole moments $\mathbf{Q}_{ij}$ with the field gradients $-\partial^2 V/\partial i \partial j$. If the z'-axis of this molecule, which is also the axis of its dipole moment, has polar angles (θ', ϕ') in the co-ordinate system of the first molecule, then the dipole–dipole interaction energy between these two fixed molecules is

$$U_{11}(\theta, \phi; \theta', \phi') = -\mu^2 r^{-3} \cos\theta'[3\cos^2\theta - 1] + \sin\theta' \sin\theta \cos(\phi - \phi'). \quad (1.16)$$

This energy has equal angular ranges of positive and negative sign. In a gas, however, the molecules are free to rotate so we must average over all relative orientations, including a Boltzmann weighting factor to take account of the greater amount of time spent in orientations of lower energy. This average then leads to an attractive interaction which, provided $\mu^2/r^3 \ll kT$, has the explicit form

$$U_{11} = -\frac{2\mu^4}{3kTr^6}. \quad (1.17)$$

The dipole–quadrupole interaction similarly treated gives an attractive potential $U_{12} = -a_{12}r^{-7}$ and the quadrupole–quadrupole interaction a potential $U_{22} = -a_{22}r^{-8}$, where a_{12} and a_{22} depend on the product of the molecular moments involved and vary inversely with temperature.

This picture of the electrostatic interaction, as far as it goes, is quite satisfactory but it must be used with caution, not only because it is not the only component of the attractive force but also because there is no guarantee that a multipole expansion like (1.15) will converge for very small intermolecular separations.

It became clear long ago that these permanent multipole forces alone are insufficient to explain observed molecular interactions. Apart from anything else, the term which they contribute to the Van der Waals equation has the wrong temperature dependence, experiment showing the need for an additional temperature-independent attractive potential. The origin of this potential, introduced by Debye (1929), is physically quite clear. In an electric field $\mathbf{E}$ the electron distribution of a molecule will be distorted and the distortion will give rise to an additional dipole moment $\boldsymbol{\mu}' = \alpha\mathbf{E}$. For a spherical molecule the polarizability α is a simple constant so that the induced dipole is parallel to the field. More generally α is a tensor so that the dipole will make an angle with $\mathbf{E}$ but will still have its major component parallel to it. From (1.15) the field due to the permanent dipole of the first molecule is $-\mu r^{-3}\cos\theta$ so the interaction energy between permanent and induced dipoles is

$$-\tfrac{1}{2}\mu'\mu r^{-3}\cos\theta = -\tfrac{1}{2}\alpha\mu^2 r^{-6}\cos^2\theta. \qquad (1.18)$$

When an extra factor of 2 is inserted to allow for the symmetrical interaction of the second molecule on the first and the result is averaged over angles, this gives an additional attractive potential

$$U' = -\tfrac{1}{3}\alpha\mu^2 r^{-6}. \qquad (1.19)$$

The correct expression with α a tensor is more complicated but has the same general form. In the same way we can calculate interaction terms with higher induced moments, all of which will give terms falling off more rapidly than r^{-6} so that at reasonable distances they are negligible.

The quantum description of the induced dipole moment is worth a moment's consideration here. Distortion of the electron wave functions can only be achieved by adding in contributions from excited states. This requires energy but, since the excited states may increase the dipole moment parallel to the field, the total energy of the system can be lowered. The components of the polarizability tensor α can be found by ordinary second-order perturbation theory and have the form

$$\alpha_{xy} = 2e^2 \sum_{i \neq 0} \frac{\langle \psi_0 | x | \psi_i \rangle \langle \psi_i | y | \psi_0 \rangle}{W_i - W_0}, \tag{1.20}$$

where W_i is the energy of the excited state ψ_i and W_0 that of the ground state ψ_0. To simplify this result it is often a good approximation to set $W_i - W_0$ for all states i equal to the ionization energy I of the molecule, because the excited states in question are generally not far removed in energy from the beginning of the ionization continuum. In a general way then, we may expect a molecule with a low ionization energy to have a relatively high polarizability. The average polarizability of the water molecule as derived from optical data is $1{\cdot}5 \times 10^{-24}$ cm^3, which is fairly low, in keeping with its moderately high ionization potential.

The sum of the two energies given by (1.17) and (1.19) still does not give the correct interaction between molecules. This is obvious, since these equations allow no attraction between molecules lacking a permanent dipole moment μ, whereas we know that the attractive force between such molecules is by no means negligible. The additional force, which is of quantum origin and universal effect, is called a dispersion force (London, 1937).

Dispersion forces arise because of the fluctuations and uncertainties which are fundamental in quantum mechanics. Thus, though the expectation value $\langle \mu \rangle$ of a molecular dipole may be zero, the instantaneous value fluctuates about zero because $\langle \mu - \langle \mu \rangle \rangle^2 = \langle \mu^2 \rangle - \langle \mu \rangle^2$ does not vanish. This instantaneous fluctuating dipole can cause polarization effects in a neighbouring molecule in the same way as does a permanent dipole and this leads to an attractive potential varying as r^{-6}. Details of the calculation are given by London (1937) and Buckingham (1965) and in standard texts on quantum mechanics and we will not repeat them here. The approximate result found by London was

$$U'' \simeq -\frac{3I\alpha^2}{4r^6}, \tag{1.21}$$

where α is the polarizability of the molecule, given by a formula like (1.20), and I is its ionization potential. When the tensor nature of α is taken into account for a non-spherical molecule then the exact expression is, of course, much more complicated.

It is also worth while to note that the ordinary derivation neglects the finite velocity of light and assumes that interactions between the two molecules are instantaneous. However, the dipoles involved in the dispersion forces fluctuate at optical frequencies so that, if the separation between the two molecules is not considerably less than an optical wavelength, there will be important phase effects which can be shown to reduce the interaction energy to an r^{-7} dependence (Longuet-Higgins, 1965). This effect will not need to concern us here.

Finally we note the existence of repulsive exchange forces which arise when the electron clouds of the two molecules begin to overlap. These occur because of the exclusion principle which forbids two electrons to occupy the same orbital unless their spins are antiparallel. Since all orbitals are already occupied by electron pairs, any overlap means that some electrons must be transferred to anti-bonding orbitals of higher energy, which implies a repulsive interaction. This interaction, being of very short range, depends upon the exact relative positions of the two molecules and it is not worth while to consider the almost endless possibilities involved until we can deal with molecules fixed in a crystal lattice.

The three most important components of the interaction energy between two water molecules, as given by (1.17), (1.19) and (1.21), all vary as r^{-6} and it is helpful to compare their magnitudes. At 293 °K the relative values calculated by London are

$$\left.\begin{array}{lr} \text{permanent dipole interaction} & 190 \\ \text{induction energy} & 10 \\ \text{dispersion energy} & 47 \end{array}\right\} \times 10^{-60}\ \text{erg cm}^6. \qquad (1.22)$$

The dispersion forces thus account for nearly 20 per cent of the interaction energy, even for a strongly polar molecule like water. We should emphasize again, however, that these values do not necessarily apply when the two molecules are virtually in contact, as in the liquid or solid, and we will need to return later to this problem.

CHAPTER 2

STRUCTURE AND ENERGY OF ORDINARY ICE

Following on our survey of the properties of the water molecule, we now come to consider the ways in which these molecules link together to form a liquid or a solid phase. This is a large subject and we shall treat it in several stages. In the present chapter we look at ordinary ice—Ice I_h—and see what is known about its crystal structure and cohesive energy. With this discussion as a background, we shall then go on to consider the other forms of ice which can exist at low temperatures or at high pressures and to take an over-all view of the phase diagram of the material called water.

2.1. Crystal structure of Ice I_h

The equilibrium structure to which a material crystallizes under given conditions of temperature and pressure is completely determined by the interaction forces between its molecules and these, in turn, are determined by the electronic structure of those molecules, as we have seen. Unfortunately it is almost impossible to proceed uniquely in this way from molecules to crystals in any but the very simplest solids, because the interaction forces are not known with sufficient accuracy. We must therefore usually content ourselves with observing that the crystal structure which actually occurs is consistent with what is known of the molecular interactions, and with comparing the value which we can calculate for its cohesive energy with that found from experiment. In the case of ice, determination of the crystal structure has itself posed very difficult problems which have only been answered by reference back to the structure and interaction of the water molecules.

Some clue as to the structure of a crystal can usually be found from observation of the habits of natural crystals, particularly when these have grown in an environment which is reasonably symmetrical. Ice crystals which have grown by direct sublimation from the vapour phase in the atmosphere are ideal for this purpose but, though they provide some clues, they end up posing more questions than they answer, as we shall see later. The forms of these snowflakes are generally familiar, as is their bewildering variety. A beautiful collection of photographs appears in the classic book by Nakaya (1954). Most of the crystals have striking hexagonal symmetry, the most commonly noted habit being a thin hexagonal plate, which may be featureless but usually has complex dendritic branches with remarkable hexagonal symmetry down to their finest details, as illustrated in plate 1. Rather less often the crystals are observed to form hexagonal prisms or long hexagonal needles, and, of course, many asymmetric crystals are also found. We shall return to discuss the significance of these different crystal habits in chapter 5. For the present we simply note that they strongly suggest a hexagonal symmetry for the molecular arrangement in the crystal, though they cannot be taken to establish this definitely since hexagonal plates can be produced from cubic structures by appropriate twinning along (111) planes, as is seen for example in the silver bromide grains of a photographic emulsion.

The standard method of determining crystal structure is by X-ray diffraction and this method was applied to ice by Bragg and others some fifty years ago. The early results were not, however, in agreement and the situation was finally clarified by Barnes (1929). X-rays are scattered primarily by the electron distribution in the crystal and, from the discussion of chapter 1, this is concentrated around the oxygen nucleus in each water molecule, though there is an appreciable electron density around the two protons. The X-ray diffraction results thus basically determine the positions of the oxygen atoms, the proton positions being much more difficult to fix.

The structure of a crystal, in terms applicable to diffraction results, is specified by giving its space group and by giving, in addition, the positions of the equivalent scattering centres in the unit cell. Each of these components in the total structure is recognized by certain sets of missing spots and variations of intensity in the diffraction pattern. A general description of the symmetry elements and notations used can be found in the *International Tables for X-ray Crystallography* (International Union of Crystallography, 1952). By these means Barnes and more recent investigators concluded that ordinary ice has a crystal structure which, at least as far as the oxygen positions are concerned, belongs to the space group denoted by the Schoenflies symbol D_{6h}^4 or by the International symbol $P6_3/mmc$. This symmetry is set out in detail in the *International Tables* (pp. 304–5). Briefly the D_{6h} part of the symbol indicates the point group involved and implies first a sixfold rotation axis (C_6), to which is added an axis of twofold rotation symmetry perpendicular to the c-axis (and hence three such axes), indicated by changing C_6 to D_6. Finally the addition of a horizontal reflexion plane is shown by the subscript h. The superscript 4 is simply numerical and specifies the particular space group (obtained by adding screw axes and translational glide planes) which can be classified under the point group D_{6h}. The International symbol lists more explicitly the multiplicity and pitch of the screw axes (6_3), and the existence of mirror planes (m) and glide planes (c), while P indicates a primitive rather than a centred structure.

For our present purposes we shall not be concerned with the detailed symmetry classification of the structure but rather with the positions of the atoms in the unit cell and relative to one another. The unit cell for a hexagonal lattice is a prism set on a rhombic base with included angle $\frac{2}{3}\pi$. In this cell there are four oxygen atoms at the points $\pm(\frac{1}{3}, \frac{2}{3}, z_0)$, $\pm(\frac{2}{3}, \frac{1}{3}, \frac{1}{2}+z_0)$, where z_0 is closely equal to $\frac{1}{16}$. These are the atomic positions for the wurtzite structure, which can be considered as formed from two interpenetrating hexagonal close-packed structures with origins $(0, 0, z_0)$ and $(0, 0, \frac{1}{2}-z_0)$ respectively. It is, in fact, the hexagonal analogue of the diamond cubic structure. An expanded view of the structure is shown in Fig. 2.1, from which two things are at once obvious: (i) the oxygen positions lie in crinkled sheets which are oriented normal to the c-axis; (ii) each oxygen position is surrounded tetrahedrally by four other oxygen positions.

We shall return to the first of these observations later when discussing the mechanical properties of ice; let us now look more closely at the second. It is, we can see, very much what we might have expected from the roughly tetrahedral shape of the water molecule. It is, in fact, very reasonable to expect the molecules to pack with fourfold co-ordination so that the proton on one molecule is close to one of the negatively charged lone-pair hybrids of a neighbouring molecule, which leads almost uniquely to either a wurtzite or a diamond cubic structure. We shall investigate the energy and stability of this arrangement later on; let us now return to the determination of the structure itself.

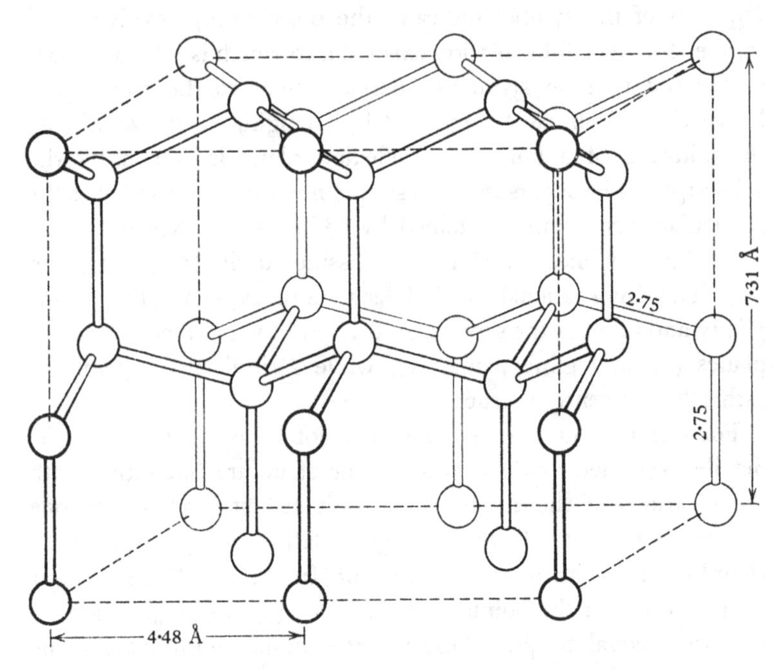

Fig. 2.1. Expanded view of oxygen atom positions in the ice structure. Note the tetrahedral co-ordination and the crinkled sheets normal to the *c*-axis. Dimensions shown in this and later figures are those for a temperature of 77 °K (Kamb, 1968). (From *Structural Chemistry and Molecular Biology*, ed. Alexander Rich and Norman Davidson. San Francisco: Freeman & Co., copyright © 1968.)

X-ray investigations, as well as determining lattice symmetries, also give precise values of lattice parameters. Measurements prior to 1958 have been surveyed, together with much related work, in papers by Lonsdale (1958) and Owston (1958) and these, together with a very extensive set of measurements by LaPlaca & Post (1960), are displayed in fig. 2.2. The X-ray value for the density of ice at 0 °C is 0·9164 g cm^{-3}, which is in excellent agreement with macroscopic measurements, the most reliable reported values ranging from 0·916 to 0·918 g cm^{-3}. We shall return to discuss the expansion coefficients deduced from these X-ray measurements in chapter 6. In passing we also note that the lattice parameters for D_2O ice, which are also discussed by Lonsdale (1958), differ by less than 0·1 per cent from those of H_2O ice over the temperature range studied.

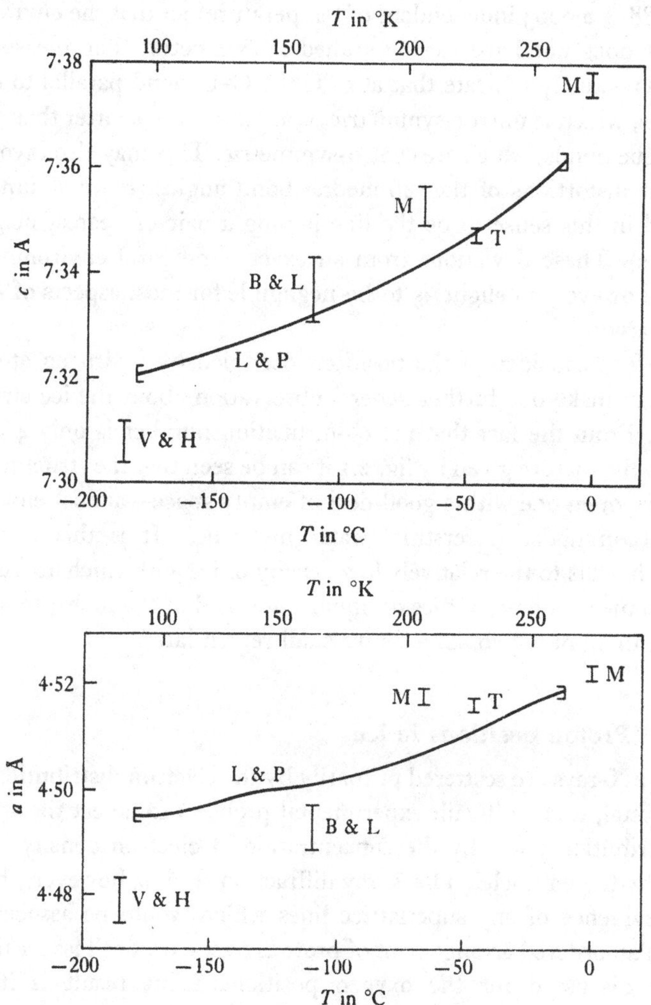

Fig. 2.2. Lattice parameters of ice as determined by diffraction methods. Isolated points measured by Megaw (M), Truby (T), Blackman and Lisgarten (B & L) and Vegard and Hillesund (V & H) are discussed by Lonsdale (1958). The continuous curve represents an extensive set of measurements by LaPlaca & Post (L & P) (1960).

Looking now more closely at the environment of each oxygen position, we note that, if this were exactly tetrahedral, then the lattice ratio c/a should have the value $\sqrt{\frac{8}{3}} = 1{\cdot}633$. The measured c/a values are rather less than this, ranging from 1·627 to 1·632 according to different observers (LaPlaca and Post find $c/a = 1{\cdot}6288 \pm 0{\cdot}0003$ independent of temperature) so that the environment does not have exact tetrahedral symmetry. The measurements actually indicate that at 0 °C the O–O bond parallel to the c-axis, which is mirror-symmetric, is about 0·01 Å shorter than the oblique bonds, which are centro-symmetric. This may also involve slight distortions of the tetrahedral bond angles (if we assume a bond in this sense to be the line joining a pair of nearest neighbours). These deviations from an exact tetrahedral environment are, however, so slight as to be negligible for most aspects of our discussion.

Before considering the positions occupied by hydrogen atoms we can make one further general observation about the ice structure. From the fact that the co-ordination number is only 4 and from the picture given by fig. 2.1 it can be seen that the structure is a very open one with a good deal of empty space—almost enough to accommodate interstitial water molecules. It is this feature which leads to the relatively low density of ice and which accounts for some of the properties of liquid water and of the high-pressure ices, to all of which subjects we shall return later.

2.2. Proton positions in ice

Since X-rays are scattered primarily by the electron distribution in a crystal, it is a difficult experimental problem to detect the small contribution made by the concentration of electron density near the hydrogen nuclei. The X-ray diffraction results, however, show no evidence of any superlattice lines which would be associated with an ordered arrangement of protons over a unit cell larger than that discussed for the oxygen positions. This result is itself sufficient to greatly limit the number of possible structures and Barnes (1929) suggested the most probable arrangement to be one with a proton at the centre of each O–O nearest-neighbour bond. Whilst such a structure is attractively symmetrical, it is unsatisfactory for several reasons. It suggests that water should be an ionic liquid with H^+ and O^{2-} ions, while in fact it consists almost entirely of undissociated H_2O molecules. It further predicts a dielectric constant for ice of the same order as for other ionic crystals, whilst the measured value (of the order of 100) suggests the presence of permanent dipoles in the structure.

Further evidence supporting the molecular nature of the ice crystal comes from studies of the infrared and Raman spectra of ice, a comprehensive review of which has been given by Ockman (1958). The spectrum at wave numbers below 1500 cm^{-1} consists of bands due to translational and rotational vibrations of water molecules (to which topics we shall return in chapter 6); in the region from 1500 cm^{-1} to 3800 cm^{-1} there are several bands, some of which can be related to the fundamental vibrations of the water molecule, and above 3800 cm^{-1} there are many overlapping bands due to overtones and combinations of the three fundamentals with one another and with lattice modes. In the fundamental region the assignments in table 2.1 can be made, the vapour values being shown for comparison, as also are the values for D_2O ice. The $\tilde{\nu}_1$ band in both ices is seen only in Raman spectra. The important point for our present concern is the relatively small shift in frequency of these vibrations in ice from those characteristics of the free molecule, suggesting that the process of binding into the solid has not greatly changed the internal structure of the molecule.

TABLE 2.1. *Wave number assignments for hexagonal ice (all in* cm^{-1}*)*

	H_2O (vapour)	H_2O (ice)	D_2O (ice)
$\tilde{\nu}_1$	3652	3143	2347
$\tilde{\nu}_2$	1595	1640	1210
$\tilde{\nu}_3$	3756	3252	2440

These considerations lead one to seek for a structure in which the integrity of water molecules as structural units is preserved. Several such structures were proposed by Bernal & Fowler (1933) based on the two principles that water molecules are preserved and that they link together so that each proton of one molecule is directed towards a lone-pair electron hybrid of a neighbouring molecule. They found a simple arrangement which, however, required a unit cell three times as large as that proposed by Barnes and which was polar. The absence of superlattice lines in X-ray patterns could perhaps be explained away in terms of sensitivity limits, but the polar structure would lead one to expect piezoelectric properties which, though they have been occasionally reported, seem now to be ruled out (Steinemann, 1953; Teichmann & Schmidt, 1965), implying that the point group involved is really D_{6h}, as stated above, rather than C_{6v}. Bernal and Fowler also discuss a non-polar ordered structure but point out that its extreme complexity and large unit cell (ninety-six molecules) make its occurrence rather unlikely. Other models, mostly of a polar nature, have been suggested elsewhere in the literature but there is now enough evidence to rule them all out for various reasons.

In concluding their discussion of possible ice models, Bernal and Fowler make the statement: 'It is quite conceivable and even likely that at temperatures just below the melting point the molecular arrangement is still partially or even largely irregular, though preserving at every point tetrahedral co-ordination and balanced dipoles. In that case ice would be crystalline only in the positions of its molecules but glass-like in their orientation.' This conclusion, reinforced by some additional evidence provided by Pauling (1935) which we shall discuss in the next section, is essentially that now accepted. The only modification is that we believe that the energy differences between most of the possible configurations are so small that, before the temperature is lowered sufficiently for them to become important, the molecular configurations are frozen and the disorder persists down to 0 °K.

A model of one of the possible disordered structures is shown in fig. 2.3. Because the H–O–H angle in a free water molecule is about 105°, which is quite close to the tetrahedral angle of 109° 28′, the protons are often drawn as though located exactly upon the O–O bond lines, as we have done in fig. 2.3. This is not, however, a necessary feature of the model, as we shall see later, though the small displacements involved in other configurations seem to have no practical implications. This 'statistical' model can be simply described in terms of the positions of the oxygen nuclei and a distribution of protons among the two possible sites on each O–O bond according to the two rules (i) there is only one proton on each bond, (ii) there are only two protons close to each oxygen nucleus. To gain direct evidence about hydrogen positions in ice it is necessary to use a diffraction technique which is more sensitive to scattering by protons and their accompanying small electron

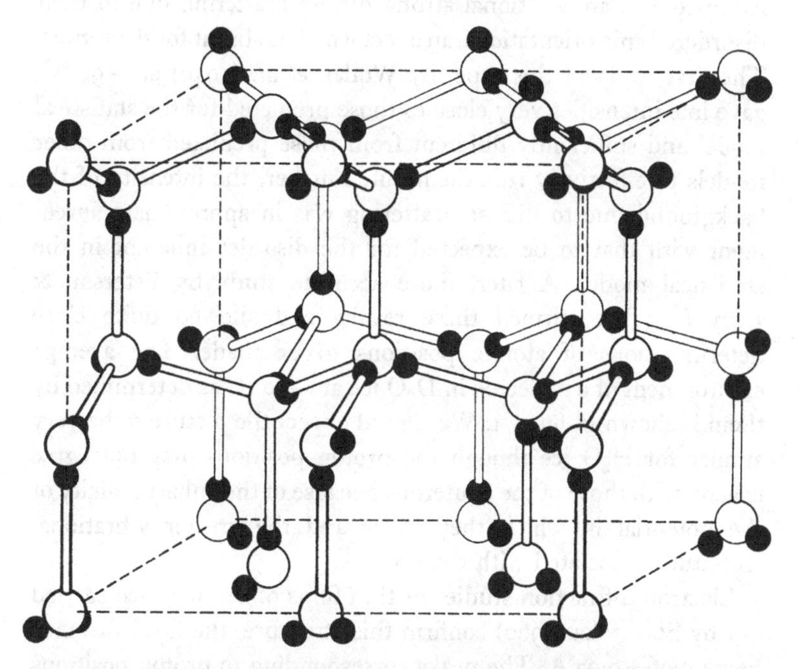

Fig. 2.3. A typical disordered arrangement of protons in the ice structure. Each oxygen has two close protons, forming an H_2O molecule, and there is one proton on each bond.

density than are ordinary X-ray methods. Two such techniques are available: electron diffraction and neutron diffraction. Before considering these in detail, let us ask what sort of results we might expect to find. In all cases the primary beam samples the diffracting centres in a considerable volume of crystal, though this is limited in depth in the case of electron diffraction by the small penetrating power of the electron beam. The diffraction pattern will thus be essentially that of the average structure seen by the beam, with, superimposed upon it, a diffuse background due to the variation of individual scattering centres from this average. The statistical structure of Bernal and Fowler and of Pauling should therefore result in a diffraction pattern with a 'half-proton' scattering centre at each of the two possible proton positions on each bond.

Neutron diffraction studies have been made with heavy ice, D_2O, rather than with ordinary ice, because the protons in ordinary ice introduce an additional strong diffuse scattering due to their disordered spin orientations, an effect which is absent for deuterons. The first study of this type, by Wollan *et al.* (1949) at −90 °C, gave line intensities very close to those predicted for the statistical model and sufficiently different from those predicted from other models to effectively rule them out. Further, the intensity of the background due to diffuse scattering was in approximate agreement with that to be expected for the disorder inherent in the statistical model. A later, more accurate study by Peterson & Levy (1957) confirmed these results and allowed quite close determinations of atomic positions to be made. The average environment of a molecule in D_2O ice at −50 °C as determined by them is shown in fig. 2.4. We should expect the picture to be very similar for H_2O ice though the proton positions may not agree exactly with those of the deuterons because of the anharmonicity of the potential in which they move and the greater vibrational amplitude associated with the protons.

Electron diffraction studies of thin films of ordinary ice carried out by Shimaoka (1960) confirm this structure, the O–H distance being $0{\cdot}96 \pm 0{\cdot}03$ Å. The peaks corresponding to proton positions were observed to be elongated in directions perpendicular to the O–H bond, which implies either that the thermal vibrations are anisotropic (which is certainly true) or that the H–O–H angles deviate slightly from the tetrahedral value, an average of the possible misalignments being seen. We shall return to this later.

Additional confirmation of the statistical model comes from experiments of a quite different nature involving nuclear magnetic resonance. The protons in the water molecule possess a nuclear magnetic moment so that their environment can be investigated by studies of the shape of their resonant absorption line in a steady applied magnetic field. In an isolated water molecule the resonance is very sharp and this is also the case in liquid water for, although the interactions with neighbouring molecules are large, they vary so rapidly in comparison with the radiofrequency resonance signal that only the average environment is detected. On the other hand, molecules fixed in the ice lattice interact strongly and essentially statically with the magnetic moments of the protons on neighbouring molecules and this interaction leads to a very broad line,

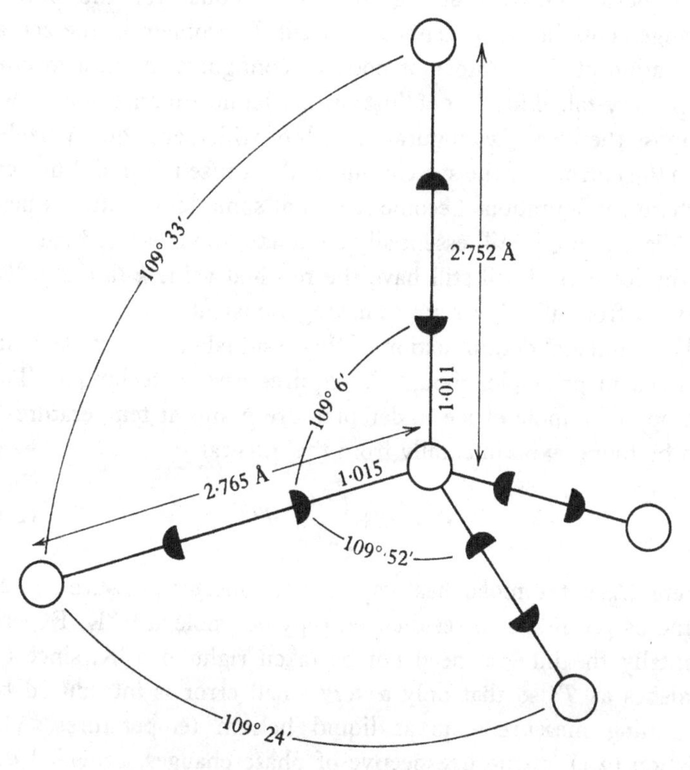

Fig. 2.4. Average environment of a molecule in D_2O ice at −50 °C as determined by neutron diffraction (after Peterson & Levy, 1957). The deuterons appear in symmetrical pairs of half-deuteron positions.

of width about 10 gauss, depending on temperature. The techniques for this study were developed by Bloembergen *et al.* (1948) and their application to the study of ice by these workers, by Kume & Hoshino (1961), Korst *et al.* (1964) and Barnaal & Lowe (1967) gives useful, though not conclusive, support to the statistical model. Unfortunately the technique, which involves comparison of observed and calculated second moments and relaxation times for the absorption, is not sufficiently sensitive to proton position to distinguish any minor variations for the Pauling model.

2.3. Residual entropy associated with proton disorder

The Bernal–Fowler–Pauling statistical model for the proton arrangements in ice presents a very subtle problem in the actual evaluation of the number of possible configurations in a macroscopic crystal. This is not just an academic exercise for, if we suppose there are ζ configurations, all of which are equally likely, then the entropy of the system due to this cause is $k \ln \zeta$. Further, if these configurations become frozen at some temperature where the disordering is still essentially complete, the measured entropy of the ice crystal will still have the residual value $k \ln \zeta$ at 0 °K, entropy from all other sources having vanished.

Experimental determination of this residual entropy is straightforward in principle, though it requires careful technique. The entropy of a mole of ice under pressure p and at temperature T can be found experimentally from the integral

$$S = S_0 + \int_0^T \frac{C_p}{T}\, dT, \tag{2.1}$$

where C_p is the molar heat capacity at constant pressure and S_0 is the as yet unknown residual entropy per mole at 0 °K. Experimentally the integral need not be taken right to 0 °K, since C_p vanishes as T^3 so that only a very small error is introduced by beginning measurements at liquid helium temperatures. The relation (2.1) is true irrespective of phase changes, provided the latent heat involved is included as a singularity of appropriate size in C_p, so that if we raise the temperature until the crystal becomes a vapour then (2.1) still gives its entropy per mole. This entropy is, however, a quantity which can be evaluated independently, provided molecular interactions in the vapour are small, from the partition function of an individual molecule. Comparison of the two results then gives the residual entropy S_0.

This approach was used by Giaque & Stout (1936), who measured the specific heat of ice down to 15 °K, extrapolating to lower temperatures to evaluate the integral as closely as possible. The extrapolation has since been confirmed by Flubacher *et al.* (1960). We shall return to the specific heat measurements in chapter 6; for the present we simply note the final result

$$S_0(\text{expt.}) = 0{\cdot}82 \pm 0{\cdot}05 \text{ cal mole}^{-1} \text{ deg}^{-1}. \tag{2.2}$$

It is this value which must be explained in terms of proton disorder in the crystal.

The number ζ of possible proton configurations in ice was first calculated by Pauling (1935), who used a very simple analysis and obtained a value for S_0 in excellent agreement with experiment, as follows. Consider an ice crystal containing N molecules, where N is sufficiently large that surface effects can be neglected. This crystal contains $2N$ O–H...O bonds, on each of which the proton has two possible positions, leading to 2^{2N} configurations. Consider, however, the four bonds made to a single molecule. These have sixteen possible configurations but only six of these satisfy the requirement that the molecule should exist as H_2O. This consideration reduces the possible number of configurations by a factor $(6/16)^N$, giving a final value $\zeta = (3/2)^N$ and an entropy

$$S_0 = Nk \ln (3/2). \tag{2.3}$$

This gives the numerical value 0·806 cal mole^{-1} deg^{-1}, which is in excellent agreement with the experimental value (2.2).

This is not the only way of obtaining this result. Pauling showed that one can proceed in the opposite sense by recognizing that there are six possible orientations for each water molecule, giving 6^N configurations, if the proper formation of bonds is neglected. The probability that a given bond is properly made from a proton and a lone pair is, however, $1/2$ and there are $2N$ bonds, so the possible number of configurations must be reduced by a factor $(1/2)^{2N}$, giving $\zeta = (3/2)^N$ once more. Another slightly different derivation, leading again to the same result, was put forward by Bjerrum (1951), but expressed this time in terms of the number of possible orientations available to a molecule which is being added to the growing face of a crystal with which it is required to make two proper bonds.

For a long time these arguments were believed to have established exactly the value $Nk \ln (3/2)$, subject only to the equal-probability assumption of the statistical model and to the neglect of surface effects. Quite recently, however, the calculation has been reconsidered from a variety of viewpoints and it is now recognized that small corrections are required to the calculated value.

A simple argument showing the origin of the correction terms and suggesting their magnitude has been given by Hollins (1964) and is most easily appreciated for the case of Pauling's second calculation, as outlined above. The essential defect in this calculation, which appears in modified form in the others, is the assumption that the probabilities of placing protons in the lattice so as to (i) complete molecules properly, and (ii) complete bonds properly, are independent. This is in fact the case for an open or 'dendritic' bonding structure, but breaks down when the effects of closed rings of bonds are taken into account.

In ice, as can be seen from fig. 2.3, there are many six-membered rings but none of smaller size. If we consider the polarities of water molecule 'vertices' around such a ring, then they must have a pattern such as + −, + −, + +, − +, − −, + −, where the commas represent the positions of the bonds and must therefore separate a pair of opposite signs. Since, however, the sequence represents a ring, its first and last signs must be opposite if the final closing bond is to be correctly made. This will be the case if the number of similar pairs, + + or − −, in the ring is even. Once the sign of the first pole has been chosen for a given molecule, the probability that the second has the same sign is 1/3, so the probability of finding an even number of similar pairs in the series, and hence of making the final bond correctly, is equal to the sum of the coefficients of even powers of x in the expansion of $(\frac{2}{3}y+\frac{1}{3}x)^6$. This sum amounts to 365/729, while the probability of finding an odd number of pairs, and hence of not making the bond correctly, is 364/729. Consideration of the existence of rings has thus increased the probability of correctly forming the final bond from the value 1/2, characteristic of dendritic structure, to 365/729 = 0·500686.... Since, as can again be seen from fig. 2.3, each molecule participates in twelve six-membered rings, the total number of such rings in a crystal with N molecules is $2N$, so that a revised estimate for the number of configurations is

$$\zeta = (3/2)^N(1+1/729)^{2N}. \tag{2.4}$$

This result is still not exact because interference between rings and the effects of rings of larger size have been neglected. It suggests, however, that Pauling's calculated entropy value may be low by an amount of the order of 1 per cent.

Di Marzio & Stillinger (1964) have developed a series approximation for calculating the number of configurations, the first term of which, as in (2.4), is the Pauling value, with subsequent terms representing corrections for different kinds of bond circuits within the lattice. Their work has been extended and the first five terms summed explicitly by Nagle (1966) to give for the residual entropy the value

$$S_0 = 0{\cdot}8145 \pm 0{\cdot}0002 \text{ cal mole}^{-1} \text{ deg}^{-1}. \tag{2.5}$$

The correction to the Pauling value is indeed about 1 per cent and the new value is in even better agreement with the experimental result (2.2).

This evidence, taken in conjunction with that from diffraction, spectral and dielectric studies, appears to have established conclusively the validity of the statistical model for the structure of Ice I_h. A few final comments are in order. We must remember that acceptance of the statistical model does not necessarily imply that all configurations have exactly the same energy, only that the energy differences involved are small compared with kT at the temperature at which the disorder becomes frozen in. Since this temperature is around 100 °K, this implies that the energy differences are considerably smaller than 0·01 eV. It is also possible to exclude certain extreme configurations such as those giving a large electric moment to a macroscopic crystal from consideration, if this is necessary, since their weight in the statistical ensemble is extremely small. Lastly we note that the statistical model does not require the protons to be located exactly upon the O–H...O bond lines. It is true that the statistical mean positions are on the bonds and the deviations from these mean positions are not large, but this does not preclude an H–O–H angle differing slightly from the exact tetrahedral value. We shall return to this point in the next section.

All these statistical arguments apply equally to D_2O ice which, as we have seen, has an essentially identical structure. The residual entropy found experimentally in this case is 0·77 cal mole^{-1} deg^{-1} (Long & Kemp, 1936).

2.4. Hydrogen bonding and cohesive energy

To determine theoretically the electronic structure of an ice crystal and from this to calculate its cohesive energy, equilibrium atomic positions and related quantities is a very complex problem which has not yet been properly solved. The standard methods used for treating other crystals (Ziman, 1964) are indeed powerful, but their power rests largely upon the perfect translational symmetry of a crystal, which demands that the electron wave functions can be written in a form satisfying the Bloch theorem. In ice the protons are disordered, as we have seen, and this destroys the exact translational symmetry of the crystal structure. The result is that, instead of the elegant theory which has been developed for perfect crystals, we must use, at least in part, the sort of treatment which is at present being developed for liquid metals, glasses and other disordered systems. No such calculations have yet been made for ice, and indeed even the band-structure problem for liquid metals is still far from complete solution. We must therefore rely upon treatments involving simpler approximations to understand the structure and cohesion of ice.

In chapter 1 we discussed the interaction between two widely separated water molecules and concluded that it could conveniently be treated in terms of (i) the interaction between the permanent electric moments of the molecules, (ii) the polarization or delocalization terms due to the distortion of the electron cloud of one molecule by the field of the other, (iii) dispersion forces having their origin in co-ordinated electronic motion in the two molecules, and (iv) the repulsion due to the action of the Pauli exclusion principle when the two electron clouds begin to overlap. Most calculations of the cohesive energy of the ice crystal amount to an evaluation of these contributions for more or less realistic models. Among these it is convenient to distinguish two different types: (1) those which treat the interaction between two water molecules in terms of a hydrogen bond and sum these bonds over the whole crystal and (2) those which treat the interactions throughout the crystal on a co-operative basis. We shall consider them in turn.

A hydrogen bond between two electronegative atoms is a structure in which the hydrogen lies roughly upon the line joining the two atoms but, because of its small size, they are able to approach quite closely together. Such bonds are common in many organic materials and also play a large part in the interaction of the water molecule with other substances. It is the uniquely small size of the hydrogen atom, partially stripped as it is of its single electron, which gives this type of bond its individuality and warrants its special study (Pimental & McClellan, 1960). From the picture of the structure of ice developed in this chapter it is clear that it can be regarded as a completely hydrogen-bonded solid, each molecule forming four hydrogen bonds to its four nearest neighbours. The O–O distance on each bond is small, about 2·76 Å, and the hydrogen position is about one third of the way along each bond so that water molecules are preserved. The protons may not lie exactly upon each of the tetrahedrally directed bonds but are very close to doing so.

Calculation of the energy of a hydrogen bond itself involves consideration of the contributions from each of the four effects noted above, and in addition some allowance must be made for the effects of the other intramolecular bonds formed by each of the oxygen atoms involved. In other words we must take some account of the fact that we are dealing with a bond between two water molecules with particular relative orientations and not with a simple O–H...O structure. Despite these difficulties, however, it is possible to construct a first-order quantum-mechanical picture of the structure of the bond and from this to derive a reasonable estimate of its energy.

As an example of this approach we consider the calculations of Weissmann & Cohan (1965*a*), who performed a self-consistent field calculation for a limited set of molecular orbitals based upon the hydrogen atom and the two oxygen atoms involved in the bond. Only the 1*s* atomic orbital on the hydrogen was considered and the 2*s* and 2*p* orbitals on each oxygen, these latter being required to take the form of appropriately weighted tetrahedral hybrids to take account of the structure of the two water molecules involved. With these assumptions and a fixed O–O distance the energy was calculated as a function of proton position on the bond. As shown in fig. 2.5 for the separation 2·76 Å appropriate to bonding in ice, there is a single asymmetric minimum for a proton position about

0·83 Å from the oxygen nucleus to whose molecule it belongs, this distance being about 0·2 Å less than that observed experimentally. The curve of fig. 2.5 maintains its general shape for all O–O distances from 2·5 to 3·0 Å and no subsidiary minimum near the second oxygen atom appears. The calculation as a whole is obviously not a very good approximation to reality, since the equilibrium O–O distance is found to be about 2·2 Å, but the method has the virtue of being based upon first principles.

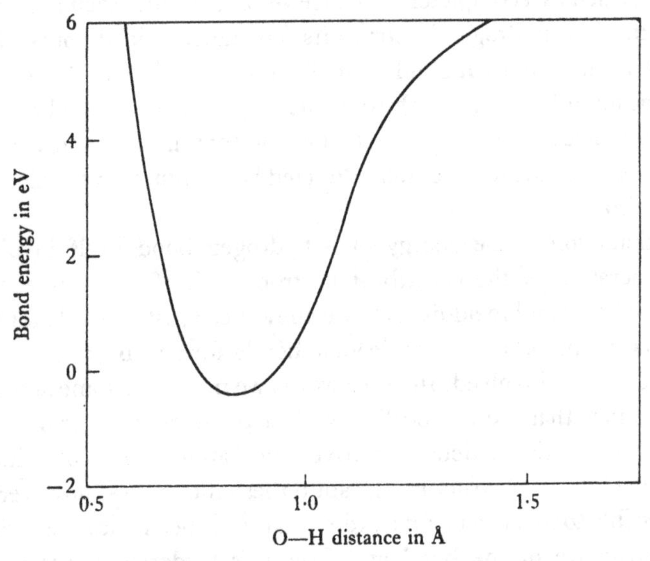

Fig. 2.5. The energy of an O–H...O bond between two water molecules as a function of proton position for a fixed O–O distance of 2·76 Å, as calculated by Weissmann & Cohan (1965*a*).

The calculated energy of the bond between two molecules with O–O distance 2·76 Å (using a better evaluation of the major three-centre integrals than was used for fig. 2.5) was 0·36 eV. This bond energy is the total interaction energy between the two molecules, constrained by the presence of their neighbours, so that, since there are just two bonds per molecule in the crystal, the lattice energy is calculated to be 16·4 kcal mole^{-1}. This neglects, of course, any interactions between molecules other than nearest neighbours, except for those taken into account by the constraints on the oxygen orbitals.

The energy calculated in this way is really a lattice energy and the cohesive energy differs slightly from this because of the difference between the zero-point vibrational energies of the molecules in the crystal and the vapour. This point has been investigated carefully by Whalley (1957) on the basis of the best available experimental data and he concludes that the true cohesive energy of ice at 0 °K is 11·3 kcal mole^{-1}, the lattice energy being greater than this by 2·06 ± 0·17 kcal mole^{-1}. This leads to an experimental lattice energy of about 13·4 ± 0·2 kcal mole^{-1} or a bond energy of 0·29 eV, values with which the calculations of Weissmann and Cohan are in moderately good agreement.

There are other experimental methods which give a measure of hydrogen-bond energy—for example by measurement of the activation energy for viscous flow in liquid water—but the energy which they yield is that required to 'break' the hydrogen bond without separating the molecules. The bond energy measured in this way is rather less well defined than the total energy calculated above and is generally about 20 per cent lower, this being roughly the contribution of the non-polar dispersion forces.

In a calculation such as that of Weissmann and Cohan it is not really meaningful to split the total binding energy into contributions from different kinds of forces. It is, however, worth while to note that the molecular orbitals characteristic of isolated molecules are considerably distorted in the process of bond formation so that the contribution of delocalization or polarization forces to the total bond energy may be quite large. The effect of this polarization is also seen in the change which it makes in the dipole moment of the molecules involved. For a free water molecule this moment was computed in the present approximation to be 1·68 debyes (experimental value 1·84 D) while that of a completely hydrogen-bonded molecule was calculated as 2·40 D, an increase of 50 per cent. We shall return to a different treatment of this effect later.

We have remarked before that the experimental evidence does not require the protons to lie exactly upon O–O bonds but only that their mean position should do so. It is possible indeed that the H–O–H angle in an individual molecule retains its vapour value of $104\frac{1}{2}°$, since this would only involve a shift of the proton by 0·04 Å normal to the bond into one of three equivalent positions for each proton (Chidambaram, 1961). On the other hand, the large distortion of the electronic wave functions during bonding destroys the appropriateness of the free-molecule H–O–H angle, and, indeed, Frank (1958) has suggested that hydrogen bonding of water molecules may be a co-operative process in which the formation of each bond strengthens those to which it is linked by reinforcing the tetrahedral hybridization of the oxygen orbitals. If this is true then the O–H–O bonds may be very closely linear, though, because of the asymmetric hydrogen positions, this need not be exactly the case.

Another quantum-mechanical approach which has been exploited by Coulson and others (Coulson, 1959) is to treat the four different energy contributions essentially separately and to regard the delocalization effects as causing a resonance between various possible ionic and covalent structures for the bond (such as O–H...O and O^-H^+...O or even O^-H...O^+ and other less likely structures). This procedure allows an estimate to be made of different contributions to the energy and gives the results shown in the fourth column of table 2.2. It is interesting to note that all the contributions are of comparable magnitude so that it is not really possible to produce a simple model for the bond.

TABLE 2.2. *Components of the lattice energy of ice (in kcal mole*$^{-1}$*) as calculated by various authors*

The experimental value is 13·4 kcal mole^{-1}.

	Bernal & Fowler (1933)	Rowlinson (1951)	Bjerrum (1951)	Coulson (1959)	Coulson & Eisenberg (1966*b*)	Campbell *et al.* (1967)
Electrostatic	14·2	9·4	14·9	12	6·5	8–9
Polarization	—	0·3	—	16	2·0	—
Dispersion	4·1 }	1·8	{ —	6	—	—
Overlap	−6·8 }		{ —	−16·8	—	—
Total	11·5	11·5	—	17·2	—	—

Before leaving discussion of the hydrogen bond itself we should mention another type of calculation which, though it throws no light on the nature of the bond, is often useful in deriving some of its properties. The heart of the method is to assume that the three nuclei O–H...O are collinear and interact with one another in a pair-wise fashion through central potential functions. These functions are written parametrically as a combination of Gaussian functions, representing simple bonds, with an over-all exponential repulsion and inverse-power-law electrostatic attraction between the oxygen atoms. The disposable parameters are then chosen so as to give agreement with a certain number of experimental data, such as the O–O distance, until they have all been fixed. The potential model then allows predictions of other quantities of physical interest to be made.

The best-known example of this approach is that of Lippincott & Schroeder (1955), who used potentials of the type described above and involving a total of ten disposable parameters. These they evaluated by assuming the same potentials to apply for isolated molecules, for which the structure is known, by requiring that the three atoms be in equilibrium at the observed O–O distance and by making estimates or arbitrary choices of the remaining parameters to give best agreement with experiment for O–H...O bonds in different materials. The computed curves of bond energy as a function of proton position thus have no fundamental basis but, through their use of empirical data, may represent a rather good approximation to reality.

Figure 2.6 shows the calculated energy as a function of proton position for an equilibrium O–O distance of 2·76 Å as in ice. Comparison with fig. 2.5 shows rough general agreement but the curve of Lippincott and Schroeder has a distinct secondary minimum at a distance of about 1 Å from the second oxygen of the bond, the primary minimum being 1·0 Å from the oxygen to which the proton belongs. This second minimum is a feature of all curves calculated by this method for equilibrium bond lengths greater than about 2·6 Å (Lippincott *et al.* 1959), though it disappears for shorter bonds. There does not at present seem to be any obvious way to establish the existence or non-existence of this second minimum but this will not need to concern us further. In some other features, however, the curves of Lippincott and Schroeder are certainly more realistic than those of Weissmann and Cohan, as indeed is to be expected from the way in which they were derived. For example, the equilibrium O–O distance calculated by Weissmann and Cohan is less than 2·2 Å while the potential of Lippincott and Schroeder is based on assumption of the correct O–O distance and gives the reasonable O–O potential shown in fig. 2.7. The O–O vibration frequency calculated for this potential is about 1×10^{12} sec^{-1}, which is close to the experimental value for the longitudinal optical mode frequency which, as we shall discuss in chapter 6, is at about $6·6 \times 10^{11}$ sec^{-1}.

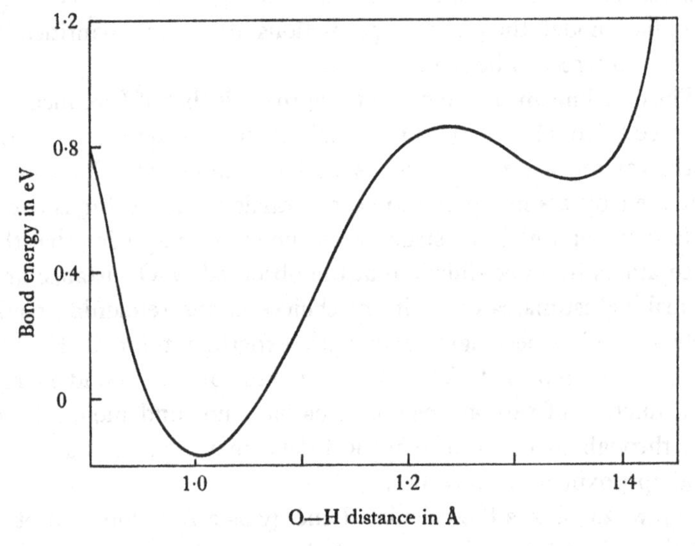

Fig. 2.6. The energy of an O–H...O bond as a function of proton position for a fixed O–O distance of 2·76 Å, as calculated from the semi-empirical model of Lippincott & Schroeder (1955).

Turning from studies of the hydrogen bond itself, there have been several rather different calculations of the cohesive energy of ice based, for the most part, on a treatment of the interaction of the permanent multipoles of the water molecules, supplemented by additional terms to account for polarization effects, dispersion forces and overlap repulsion. One of the first of these calculations was made by Bernal & Fowler (1933) using a three-charge model for the water molecule and making very simple estimates of the other energies involved. The agreement of the calculated value with experiment was reasonable, but this was, as they pointed out, largely fortuitous. The actual calculated values are given in table 2.2.

A much more careful calculation was carried out by Rowlinson (1951) using first a three-charge and then a four-charge model whose parameters were deduced from the interaction potential of water molecules as expressed by the second virial coefficient of water vapour. The four-charge model so deduced had charges of $+0{\cdot}32e$ on each of the protons and two similar negative charges

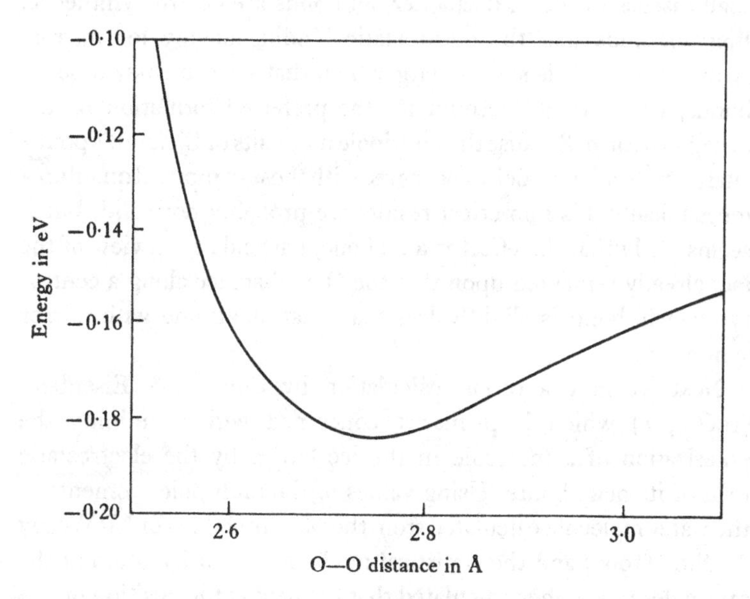

Fig. 2.7. The effective interaction potential between two oxygen atoms joined by a hydrogen bond O–H...O as calculated from the model of Lippincott & Schroeder (1955).

separated by 0·5 Å on a line through the oxygen nucleus and perpendicular to the HOH plane. As we remarked in chapter 1, this model gives the correct dipole moment for the molecule but its quadrupole moments differ in sign and magnitude from those derived from a quantum-mechanical treatment. However, the final computed lattice energy is in moderate agreement with experiment, as shown in table 2.2, though the individual components differ considerably from the estimates of Bernal and Fowler or of Bjerrum (1951), who calculated only the electrostatic component for a molecular model consisting of four charges $\pm 0{\cdot}171e$ at the vertices of a regular tetrahedron. Bjerrum's calculation is, however, of special interest, since he distinguishes between the bonding energies for two molecules which are in mirror-symmetric orientations about the bond joining them and two which are in centro-symmetric orientations. In the Ice I_h structure one-quarter of the bonds are mirror-symmetric and the remainder centro-symmetric, while in the diamond cubic analogue, Ice I_c, which we shall discuss in the next chapter, all bonds are centro-symmetric. Bjerrum finds that the electrostatic binding energy for mirror-symmetric bonds is slightly larger than that for centro-symmetric bonds, which would account for the preferred formation of the hexagonal form. Because the multipole moments of Bjerrum's point-charge molecular model do not agree with those computed quantum-mechanically, his numerical results are probably not valid, but it seems likely that the effect is a real one, particularly in view of the fact already remarked upon that the O–O distance along a centro-symmetric bond is slightly less than that along one with mirror symmetry.

Next we note a recent calculation by Coulson & Eisenberg (1966*a*, *b*) which is primarily concerned with examining the polarization of a molecule in the ice lattice by the electrostatic field of its neighbours. Using values of the multipole moments of the water molecule calculated from the wave functions of McWeeny & Ohno (1960) and the statistical model for the orientation of the molecules in ice, they calculated that the field at the position of one molecule due to the permanent multipole moments of all other molecules in the crystal is $0{\cdot}385 \times 10^6$ e.s.u. cm^{-2}. A notable feature of this result is that over 20 per cent of the total field is contributed by the quadrupole moments of nearest neighbours and just 20 per cent of the total is contributed by the dipole moments of second and further neighbours. This field is in the direction of the permanent dipole moment of the molecule concerned and so increases its moment by a polarization effect; the quadrupole and higher moments are probably changed too but this has not been calculated. Such an increase in dipole moment is, of course, a co-operative effect among all the molecules and, taking the polarizability as $1{\cdot}59 \times 10^{-24}$ cm^3, the water molecule dipole moment is increased from the value $1{\cdot}78$ D assumed initially to $2{\cdot}60$ D. This increase is very close to that computed by Weissmann & Cohan

(1965*a*) from a very different starting point, as we discussed earlier. The new value of the field resulting from these enhanced moments is $0{\cdot}516 \times 10^6$ e.s.u. cm^{-2}. Because of the statistical nature of the crystal, both these quantities are average values and there is considerable fluctuation from one molecular position to another.

The electrostatic and induction energies calculated in this way are shown in table 2.2, from which it is clear that their sum is considerably smaller than that deduced from any of the less realistic point-charge models. Coulson and Eisenberg did not calculate a dispersion or repulsion energy so no total energy is given but it is clear that these terms must contribute a greater binding effect than estimated by either Bernal and Fowler or Rowlinson if the total lattice energy is to be approximately right. A pointer in this direction is given by Kamb (1965*b*) who, from a study of the energy of the high-pressure Ice VII whose cohesion is dominated by dispersion forces, concluded that these must have about 1·5 times the strength given by the usual London formula in the case of water molecules. This gives some hope that this more careful calculation may yield a satisfactory value for the binding energy.

Finally we note the extensive computer calculation by Campbell *et al.* (1967) in which multipole energies were evaluated exactly for rather large groups of molecules instead of taking average interactions for the more distant members. Interactions up to quadrupole–quadrupole were included for all molecules, using multipole moments derived by Glaeser & Coulson (1965), and the result was as shown in the last column of table 2.2. Because of the exact nature of the electrostatic calculation, given the assumed multipole moments, it was possible to evaluate the electrostatic energy differences between different molecular configurations. These were typically a few per cent of the total energy and so amounted to about 10^{-2} eV per molecule. This is rather larger than allowed by the Pauling model but the calculation represents, of course, only one part of the total energy and these differences might well be reduced greatly in a complete treatment.

To summarize this section we can only say that, whilst we have at present a fairly good qualitative idea of the nature of hydrogen bonds and the cohesion of ice crystals, the various estimates of the energies involved are only in rather rough agreement with one another. It is true that much of the disagreement may be due to ambiguity in the definitions of the various components of the energy but the apparent good agreement between calculated and experimental values of lattice energy appears to be largely fortuitous. Whilst it is unlikely that there will be any revolution in present views of the nature of the bonding forces involved, there is clearly a need for a considerable amount of basic theoretical refinement in the calculations of cohesive energy.

CHAPTER 3

OTHER FORMS OF ICE

All materials, even completely pure ones, can exist in a variety of physical forms, the stable form in any situation depending upon the temperature and pressure. The relation between the different possible forms is expressed thermodynamically through the Gibbs free energy,

$$G = U - TS + pV, \tag{3.1}$$

and the condition that, for a system in equilibrium at a specified temperature and pressure, G is a minimum. Thus at low temperatures and pressures the internal energy U is the main factor determining stability and the equilibrium form is that in which bonds between neighbouring molecules are formed in an optimum way. At higher temperatures the entropy S begins to play an important role and a disordered liquid structure is more stable than a crystal, despite the fact that its energy is higher because intermolecular bonds are not so well made. At still higher temperatures and low pressures the completely random vapour state is stable. If the pressure is not low, however, the pV term becomes increasingly important and the equilibrium shifts towards those structures which have the lowest possible volume.

In ice, as we have seen, the bonding forces between molecules are quite strongly directional and the resulting crystal structure is very open, having a molecular co-ordination of only 4 compared with the value 12 characteristic of close-packed spheres. We should thus expect that, as well as the solid, liquid and vapour forms characteristic of low pressures, there should be at least one high-pressure form which is very much more closely packed than is ordinary ice. In fact, because of the directional character of hydrogen bonds and the shape of the water molecule, there are several different high-pressure ices which tend to be successively more closely packed as their stability regime moves to higher pressures. There are also several metastable ice structures which correspond to conditional minima in the free energy G and whose stability is sufficient to warrant their separate study. In the present chapter we shall survey what is known of these different crystal forms, leaving some of their properties for treatment later.

3.1. The phase diagram of ice

Research on the properties of materials at high pressures has now been going on for something like a hundred years and reviews of its history have been given by Bridgman (1952) and by Bradley (1963). Early experiments were largely concerned with the compressibility of liquids and with the behaviour of compressed gases and here the names of Andrews and Amagat from the late nineteenth century are well remembered. Work on the high-pressure forms of ice began with Tammann, who found Ices II and III, and was continued in the classic work of Bridgman, who was the father of modern high-pressure physics during the first half of the present century and discovered Ices V, VI and VII and the metastable Ice IV. More recently, knowledge of the structures and properties of the high-pressure ices has been greatly extended by the work of Kamb, Whalley and their collaborators in the United States and Canada respectively.

The experimentally determined phase diagram for the ice/water system is shown in fig. 3.1 which is a composite of the results of Bridgman (1912, 1937), Pistorius *et al.* (1963), Brown & Whalley (1966) and Whalley *et al.* (1968). (The unit of pressure, the bar, is 10^6 dyn cm^{-1} or almost the same as 1 atmosphere or 1 kg cm^{-2}.) In addition Bridgman (1937) has followed the melting curve of Ice VII to 44 kbar and Pistorius *et al.* to 200 kbar, where the melting point is about 440 °C, without finding any sign of a new solid phase. The results of different investigators are not in exact agreement but deviations amount to only about 2 per cent in pressure over the range shown in fig. 3.1. At higher pressures, however, the uncertainties are much greater (Whalley *et al.* 1966).

The phase diagram as shown has not been completely explored and for this reason some of the phase boundaries are shown as broken lines. Some of the phases can also be produced metastably in regions where some other phase is the stable form and such extensions are shown as dotted lines. Thus, for example, if Ice III

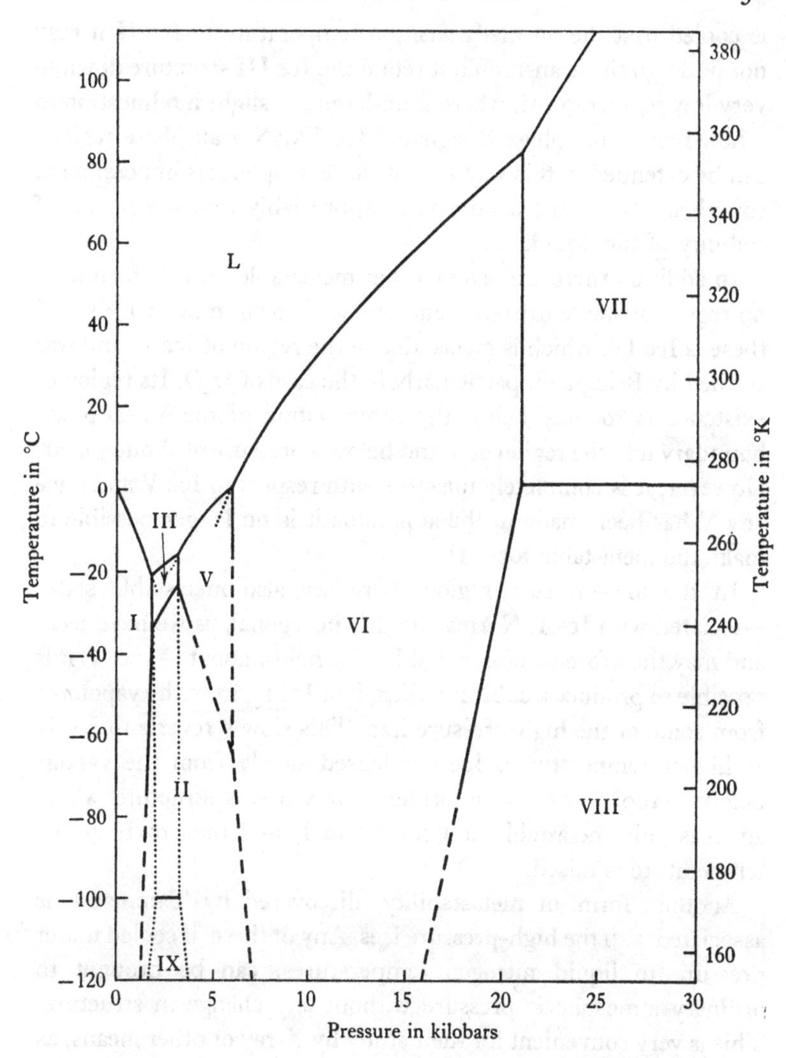

Fig. 3.1. The phase diagram of ice. Broken lines represent presumed phase boundaries which have not yet been fully investigated. Dotted lines represent metastable continuations of one phase into a neighbouring region. Ice IV, which is metastable in the region of stability of Ice V, and Ice I_c, which is metastable in the low-temperature stability region of Ice I_h, are not shown. The L–VII phase boundary continues unbroken up to at least 200 kbar, where the melting point is about 440 °C.

is cooled past the normal transition temperature to Ice II it may not undergo this transition but retain the Ice III structure down to very low temperatures, where it undergoes a slight modification to a new metastable phase designated Ice IX. Not all phase regions can be extended in this way so that, for example, it is impossible to superheat any of the solid phases appreciably into the region of stability of the liquid.

In addition there are several other metastable forms which have no region of stable existence and are not shown in fig. 3.1. One of these is Ice IV, which is metastable in the region of Ice V, and was studied by Bridgman, particularly in the case of D_2O. Its region of existence is roughly below the prolongation of the VI–L phase boundary into the region of V and below a pressure of about 5 kbar. However, it is completely unstable with respect to Ice V and once any V has been made in the apparatus it is no longer possible to make the metastable form IV.

In the low-pressure region there are also metastable states associated with Ice I. Normal Ice I is hexagonal, as we have seen, and may therefore be designated Ice I_h. Below about −120 °C it is possible to produce a cubic modification, Ice I_c, from the vapour or from some of the high-pressure ices. This slowly reverts to Ice I_h at higher temperatures. Ice condensed slowly from the vapour below −160 °C has an amorphous or vitreous structure which again is only metastable and reverts to I_c and then to I_h as the temperature is raised.

Another form of metastability, discovered by Tammann, is associated with the high-pressure ices. Any of these, if cooled under pressure to liquid nitrogen temperatures, can be brought to ordinary atmospheric pressure without any change in structure. This is very convenient for their study by X-ray or other means, as we shall see. Ices II, III, V, VI and VII, when treated in this way and then heated for a short time to about −120 °C, transform to Ice I_c, which represents a convenient method for making large quantities of this particular form (Bertie *et al.* 1963, 1964).

In determination of a phase diagram the measurements which are made are those of pressure, temperature and sample volume. It is not generally possible to measure the heats associated with transitions because the massive nature of the pressure apparatus makes thermal insulation virtually impossible. It is, however, possible to derive the heat and entropy changes involved in any transition by a simple application of thermodynamics.

When two phases are in equilibrium they are both represented by a point with co-ordinates (p, T) on a phase boundary line of the phase diagram. At such a point the Gibbs free energies of the two phases must be equal, $G_1 = G_2$, and they will remain equal if conditions move to a nighbouring point on the phase boundary. Hence

$$\left(\frac{\partial G_1}{\partial T}\right)_{\mathrm{p}} dT + \left(\frac{\partial G_1}{\partial p}\right)_{\mathrm{T}} dp = \left(\frac{\partial G_2}{\partial T}\right)_{\mathrm{p}} dT + \left(\frac{\partial G_2}{\partial p}\right)_{\mathrm{T}} dp. \quad (3.2)$$

But $(\partial G/\partial T)_{\mathrm{p}}$ is just the entropy S and $(\partial G/\partial p)_{\mathrm{T}}$ is the volume V, so that (3.2) becomes

$$\frac{dp}{dT} = \frac{S_2 - S_1}{V_2 - V_1}. \quad (3.3)$$

This is the Clausius–Clapeyron equation. From it, the slope (dp/dT) of the phase boundary and the observed volume difference between the two phases, the entropy of the transition and hence its latent heat can be found. These quantities, evaluated at the various triple points, are shown for ordinary water in table 3.1.

The phase diagram for heavy water, D_2O, was investigated up to 9 kbar by Bridgman (1935) and found to be very similar to that of ordinary water. The triple points, details of which are given in table 3.2, all occur at temperatures about 3 deg C higher than those for H_2O and the low-pressure triple points are shifted to slightly higher pressures. The volume and entropy changes have not been listed but they are of the same general magnitudes as those given for ordinary water. Ice IV is rather more clearly defined in the D_2O system, but is still unstable with respect to Ice V. These similarities are, of course, not surprising. The D_2O molecule differs from the H_2O molecule only through the added mass of the deuterons, at least as far as the present discussion is concerned. This added mass causes a shift in the frequencies, amplitudes and zero-point energies of the O–H stretching vibrations and the H–O–H bending mode. The translational vibrations of the molecule as a whole are little affected because the mass change is only from 18 to 20. We shall take up these questions of lattice

TABLE 3.1. *Triple-point relations for the phase diagram of water, H_2O*

Phases	Pressure (bar)	Temperature (°C)	Volume change (cm^3 $mole^{-1}$)	Latent heat (cal $mole^{-1}$)	Entropy change (R)
		L–I–Vap[4]			
I → L	0·006	+0·01	−1·63	1430	2·6
I → Vap			—	12160	22·4
L → Vap			—	10730	19·8
		L–I–III[1]			
I → L	2070	−22·0	−2·43	1010	2·03
III → L			0·84	916	1·84
I → III			3·27	94	0·19
		L–III–V[1]			
III → L	3460	−17·0	0·43	1103	2·17
V → L			1·42	1120	2·20
III → V			−0·98	−17	−0·03
		L–V–VI[1]			
V → L	6250	+0·16	0·95	1260	2·32
VI → L			1·65	1264	2·33
V → VI			−0·70	−4	−0·01
		L–VI–VII[2]			
VI → L	21500	+81·6	0·59	1520	2·16
VII → L			1·64	1520	2·16
VI → VII			−1·05	0	0·00
		I–II–III[1]			
I → II	2130	−34·7	−3·92	−180	−0·38
I → III			−3·53	40	0·08
II → III			0·39	220	0·46
		II–III–V[1]			
II → III	3440	−24·3	0·26	304	0·61
II → V			−0·72	288	0·58
III → V			−0·98	−16	−0·03
		VI–VII–VIII[3]			
VI → VII	21200	+5	−1·0(3)	−22	−0·04
VI → VIII			−1·0(3)	−286	−0·51
VII → VIII			0·00	−264	−0·47

(1) Bridgman (1935).
(2) Bridgman (1937) with corrected pressure (Pistorius *et al.* 1963; Brown & Whalley, 1966).
(3) Brown & Whalley (1966), Whalley *et al.* (1966).
(4) *International Critical Tables.*

dynamics again in chapter 6. For the present it is enough to note that all these effects are in the direction of slightly increasing the normal stability of a given D_2O ice structure over its H_2O counterpart, as is observed.

Before going on to a more detailed discussion of the various forms of ice and their structures it is appropriate to make one or two general observations which can be seen directly from the phase diagram.

TABLE 3.2. *Triple points for the phase diagram of heavy water, D_2O (Bridgman, 1935)*

Phases	Pressure (bar)	Temperature (°C)
L–I–Vap	0·007	+3·8
L–I–III	2200	−18·75
L–III–V	3480	−14·5
L–IV–VI	5300	−6·2
L–V–VI	6270	+2·6
I–II–III	2240	−31·0
II–III–IV	3460	−21·5

Ice I is the only ice form which is less dense than liquid water. This follows immediately from the downward slope of the phase boundary with increasing pressure and from (3.3) using the fact that the entropy of fusion is always a positive quantity because of the disordered nature of the liquid state. The fact that ice is less dense than water is, of course, of considerable geophysical and biological significance. It causes lakes to freeze from the top rather than the bottom (though this is partly due to the related density maximum in liquid water at 4 °C) and thus prevents them being completely frozen, to the extinction of freshwater aquatic life. The additional fact that ice near 0 °C can be melted by the application of pressure is also of importance for some of its mechanical properties, as we shall see in chapter 8. The high-pressure ices are not of geophysical significance on this planet because pressures in excess of 2 kbar only occur in association with rather high temperatures.

From the final column of table 3.1, giving the entropy differences between the various phases at their triple points, another interesting thing can be seen. The entropy change on melting is in all cases a little more than $2R$. By contrast, the entropy change in going from one solid form to another is very small for all transitions except those involving Ice II or Ice VIII, and possibly the I–III transition. Examination of the table entries shows that both Ice II and Ice VIII are in entropy wells with respect to surrounding phases, the well depth being about $\frac{1}{2}R$. Since we know that the protons in Ice I_h are disordered and have a configurational entropy of $R \ln (3/2) \simeq 0{\cdot}4R$, it is tempting to conclude that the protons in Ices II and VIII are ordered whilst those of all the other ices are substantially disordered. This is, in fact, true, as can be seen from dielectric measurements (Wilson *et al.* 1965; Whalley *et al.* 1966). The static dielectric constants of Ices I, III, V, and VII are all in the range 100–200, as shown in table 3.3, while Ices II, VIII and IX show no dielectric dispersion effects and have dielectric constants less than about 5, indicating that their molecular dipoles are orientationally immobile in contrast with the freedom present in the other ices.

TABLE 3.3. *Static dielectric constants of different ices* (*Wilson* et al. 1965, *Whalley* et al. 1966, 1968)

Ice	Pressure (kbar)	Temperature (°C)	Dielectric constant
I	0	−30	99
II	2·3	−30	3·66
III	3	−30	117
V	5	−30	144
VI	8	−30	193
VII	> 21	+22	≃ 150
VIII	> 21	0	Small
IX	2·3	−110	≃ 4

The crystal structures of most of the ices have now been determined with reasonable certainty and will be discussed in turn with some detail. For convenience this information is assembled in table 3.4, all the data being reduced to atmospheric pressure and a temperature of −175 °C. The structures of some of the lower ices bear a close resemblance to the allotropic forms of quartz while the highest-pressure forms have some analogy with the clathrate cages formed by water molecules about inert gas atoms.

TABLE 3.4. *X-ray structural data for ice polymorphs at 1 bar and −175 °C (Kamb, 1965a)*

Ice	Density (g/cm³)	Crystal system	Space group	Cell dimensions (Å)	Molecules in unit cell
I_h	0·92	Hexagonal	$P6_3/mmc$	$a = 4{\cdot}48$, $c = 7{\cdot}31$	4
I_c	0·92	Cubic	$Fd3m$	$a = 6{\cdot}35$	8
II	1·17	Rhombohedral	$R\bar{3}$	$a = 7{\cdot}78$, $\alpha = 113{\cdot}1°$	12
III, IX	1·14	Tetragonal	$P4_12_12$	$a = 6{\cdot}79$, $c = 6{\cdot}79$	12
IV	1·28	—	—	—	—
V	1·23	Monoclinic	$A2/a$	$a = 9{\cdot}22$, $b = 7{\cdot}54$, $c = 10{\cdot}35$, $\beta = 109{\cdot}2°$	28
VI	1·31	Tetragonal	$P4_2/mmc$	$a = 6{\cdot}27$, $c = 5{\cdot}79$	10
VII, VIII	1·50	Cubic	$Pn3m$	$a = 3{\cdot}41$	2

3.2. Ice I_c

Crystals of simple metals are often found to have either the face-centred cubic or the hexagonal close-packed structure, each of which represents one possible way of packing identical spheres as closely as possible. The (111) planes of the cubic structure are identical with the (0001) basal planes of the hexagonal and the two structures differ only through the stacking sequence of these planes, which is *ABCABC*... for the cubic and *ABAB*... for the hexagonal form. Which structure a given material will have is determined by a delicate balance of effects of higher order than simple nearest-neighbour interactions.

We have seen that ordinary Ice I_h has a crystal structure, for oxygen positions, which can be considered as two interpenetrating hexagonal close-packed structures. This is the structure taken by the form of ZnS called wurtzite, though in this case one of the sub-lattices is occupied by zinc atoms and the other by sulphur atoms. There is a cubic analogue of this structure consisting of two interpenetrating face-centred cubic lattices based on the points (0, 0, 0) and (1/4, 1/4, 1/4), which is actually the cubic structure taken by diamond, silicon and germanium. The ZnS mineral sphalerite also has this structure, which emphasizes the similarity in energy between the two forms.

It is reasonable to expect, therefore, that there may be a form of ice based upon the diamond structure, though nothing can be said from our basic knowledge about its relative stability. Such a form was found by König in thin films of ice deposited from the vapour in an electron diffraction apparatus below about -100 °C. Since that time a considerable amount of similar but more refined work has been done, as reviewed by Blackman & Lisgarten (1958), and the nature of the phase has been well confirmed: the lattice parameter is $6{\cdot}350 \pm 0{\cdot}008$ Å at -130 °C, which gives essentially the same volume per molecule as in the hexagonal phase. In fact the nearest-neighbour environment in both cases is identical, differences in structure only affecting second or more distant neighbours, as is shown in fig. 3.2. The infrared spectra of Ices I_h and I_c over the range of 350–4000 cm^{-1} are essentially identical (Bertie & Whalley, 1964*a*) and both show broadening features in the case of H_2O–D_2O mixtures which can be interpreted as due to proton disorder. Electron diffraction can also give direct evidence about the positions of the protons in the ice structure and this was taken up for Ice I_c by Honjo & Shimaoka (1957) and by Shimaoka (1960), who found an unambiguous half-hydrogen distribution with an O–H distance of 0·97 Å. We therefore conclude that Ice I_c has the same configurational entropy as Ice I_h and is in every respect simply its cubic analogue.

It seems likely that Ice I_c is metastable with respect to Ice I_h, rather than representing a new stable phase. Ordinary ice, though cooled to temperatures where the cubic form deposits from the vapour, shows no tendency to transform to it. The cubic form, however, changes readily to the hexagonal when the temperature is raised above about -80 °C. The transition is not sharp, as with a normal phase change, but proceeds at a rate which depends upon temperature (Dowell & Rinfret, 1960) and upon the previous history of the sample. Most of these studies were made with thin deposits grown from the vapour, but Bertie *et al.* (1963) were able to make reasonably large samples of cubic ice by first preparing Ice III, cooling to liquid nitrogen temperature, releasing the pressure and warming the specimen to -116 °C for a minute or so. The cubic-to-hexagonal transformation rate at -102 °C varied by more than an order of magnitude, depending upon the thermal history of the sample, which implies that some sort of nucleation process may be involved in the transformation. The latent heat

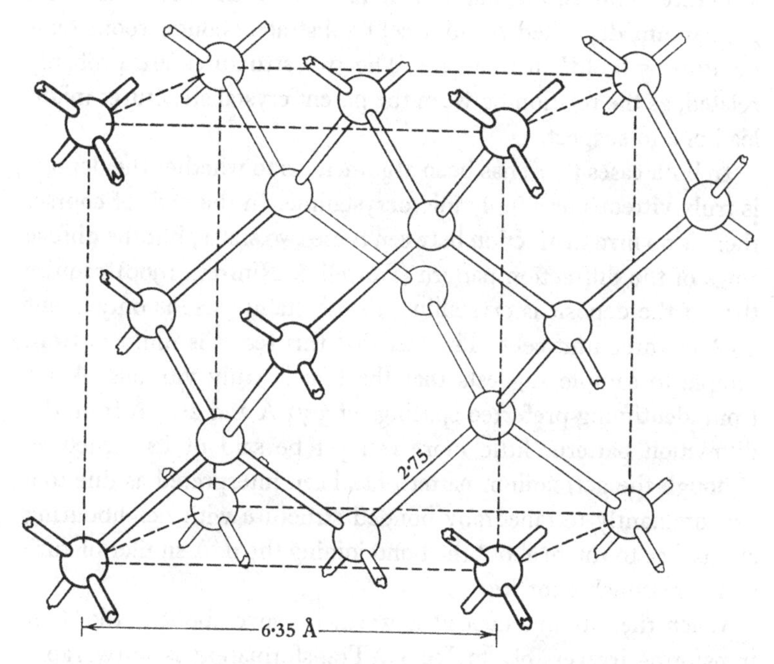

Fig. 3.2. The structure of Ice I_c, which is the cubic analogue of ordinary Ice I_h. Proton positions are not shown, but are randomly disordered (Kamb, 1968). (From *Structural Chemistry and Molecular Biology*, ed. Alexander Rich and Norman Davidson. San Francisco: Freeman & Co., copyright © 1968.)

involved is apparently very small and has not yet been measured, though Beaumont *et al.* (1961) were able to place on it an upper limit of about 1·5 cal g^{-1}.

While these results do not prove that Ice I_c is only metastable with respect to I_h at all temperatures, this is certainly the case near -100 °C, so that the region where Ice I_c is the stable form, if such a region exists, must be at a temperature well below -100 °C, the transformation from I_h to I_c then being effectively prevented by the slow rate of the molecular processes involved.

3.3. Vitreous ice

When water condenses from the vapour on to a surface at a temperature below about −160 °C a clear deposit of ice-like material is formed which apparently has an amorphous or glassy structure. This is, indeed, a phenomenon which occurs also with germanium deposited on to a cold substrate, though room temperature is 'cold' in this case. The two structures are probably related, as the relation between the parent crystal structures might lead one to suspect.

In both cases there has been argument as to whether the deposit is truly vitreous or simply microcrystalline. In the end, of course, there is no firm distinction between these two states, but the diffuse rings of the diffraction pattern (Dowell & Rinfret, 1960) require that, if the deposit is crystalline, the crystallite size is only about 20 Å or three unit cells. The fact that this result is uniform from sample to sample suggests that the film is truly vitreous. Apart from identifying preferred spacings of 3·71 Å and 2·15 Å from the diffraction pattern, little more can yet be said of its structure, although the germanium pattern has been interpreted as due to a predominantly tetrahedrally bonded structure with neighbouring atoms free to rotate about the bond joining them. A similar picture seems reasonable for ice.

When the vitreous deposit is warmed above about −160 °C it transforms irreversibly to Ice I_c. Transformation is fairly rapid above −135 °C so that at warming rates of a few degrees per minute the transformation appears to take place at about −130 °C (Dowell & Rinfret, 1960). The heat evolved in the transition seems to be 7–24 cal g^{-1} (Pryde & Jones, 1952; Ghormley, 1956, 1968; Beaumont *et al.* 1961) with the most probable value lying near the top of this range.

3.4. Ice II

When we come to consider the high-pressure ices there is much less information available than for ordinary ice and a complete description of their structure and behaviour is not possible. We must therefore be content with a rather sketchy outline. Earlier in the present chapter we pointed out that Ice II is in an entropy well of about $\frac{1}{2}R$ relative to neighbouring ices. In fact the entropy difference between Ice II and Ice I_h over the temperature range -34 °C to -75 °C, as measured by Bridgman (1912), is 0·77 cal mole^{-1} deg^{-1}, which is close to the residual entropy of proton disorder in Ice I, 0·82 cal mole^{-1} deg^{-1}, as discussed in chapter 2. This is strong evidence for an ordered proton arrangement in Ice II and this is supported by dielectric measurements (Wilson *et al.* 1965; Whalley *et al.* 1968) which, as previously discussed, show no dispersion but a simple high-frequency value of 3·66 at -30 °C and 2·3 kbar.

If Ice II is brought to atmospheric pressure at liquid nitrogen temperatures then there appears to be no change in its structure or properties and several different measurements can be conveniently made upon it. The infrared spectrum of Ice II over the range 350–4000 cm^{-1} has been measured in this way by Bertie & Whalley (1964*b*) and compared with the spectra of the other ices. The Raman spectrum has been similarly examined by Taylor & Whalley (1964). In general features the spectra of Ice II made from pure H_2O or D_2O are very similar to those of the other ices, the frequency shifts in the absorption bands away from those of Ice I_h being small compared with the shift between I_h and the vapour. From this it can be concluded that Ice II is essentially fully hydrogen bonded and probably four-coordinated, with water molecules retaining their individuality. Indeed this conclusion is reached for all the high-pressure ices.

If, however, a small amount of D_2O is included in a crystal made from H_2O (or vice versa) then an interesting effect occurs. Because of proton or deuteron mobility the added deuterons rapidly diffuse to form HDO molecules and, since the O–D stretching vibration frequencies are very different from those of O–H, there is very little coupling between them. An examination of the O–D vibration band in a crystal which is mostly H_2O therefore gives an indication of the sort of environment found by an isolated deuteron. In most ices the O–D bands found in this way are fairly broad and featureless, reflecting the continuous variety of average environments created by the proton disorder. In Ice II, however, the O–D stretching band shows four clearly resolved maxima which indicate that, instead of a statistical environment, there are four different sorts of O–D...O bonds. This implies order among the proton arrangements and allows some deductions to be made about crystal symmetry (Bertie & Whalley, 1964*b*).

More direct evidence of crystal symmetry and structure can, of course, be obtained by X-ray diffraction techniques and these have been applied to Ice II crystals at atmospheric pressure by Bertie *et al.* (1963) and by Kamb (1964). These diffraction studies have been performed both on powdered samples and on reasonably large single crystals which were found to occur in the specimens as prepared and are in substantial agreement. On the basis of the diffraction pattern Kamb has proposed a crystal structure for Ice II. The unit cell of this structure contains twelve molecules and is rhombohedral with

$$a = 7{\cdot}78 \pm 0{\cdot}01\ \text{Å}, \quad \alpha = 113{\cdot}1 \pm 0{\cdot}2^\circ. \tag{3.4}$$

The corresponding triply primitive hexagonal cell has

$$a = 12{\cdot}97 \pm 0{\cdot}02\ \text{Å}, \quad c = 6{\cdot}25 \pm 0{\cdot}01\ \text{Å}. \tag{3.5}$$

The space group is $R\bar{3}$ and the density at 1 bar and -150 °C is $1{\cdot}17$ g cm^{-3}. The Ice II structure contains the same six-membered rings of water molecules which are characteristic of the Ice I_h structure but these rings are linked together in a different way so as to form the more closely packed structure shown in fig. 3.3. In addition the rings are slightly distorted so that, though each molecule is hydrogen-bonded to just four nearest neighbours, the bond lengths and bond angles are not all equivalent. There are two types of six-membered rings in the structure and a total of just four different O–O bond lengths, ranging from 2·75 Å to 2·84 Å, in general agreement with the infrared data. The O–O–O angles are, however, quite far from tetrahedral, ranging from 80° to 129°, and it is these variations which are associated with proton order. It is certain that the H–O–H bond angle has a preferred value, though this may depend also upon the environment; it is also certain that the configuration of lowest energy for a hydrogen bond is that with the proton lying directly on the bond. These two considerations combine to make certain proton configurations energetically more favoured than others and the energy differences are apparently large enough that protons are completely ordered over the normal range of stability of the phase.

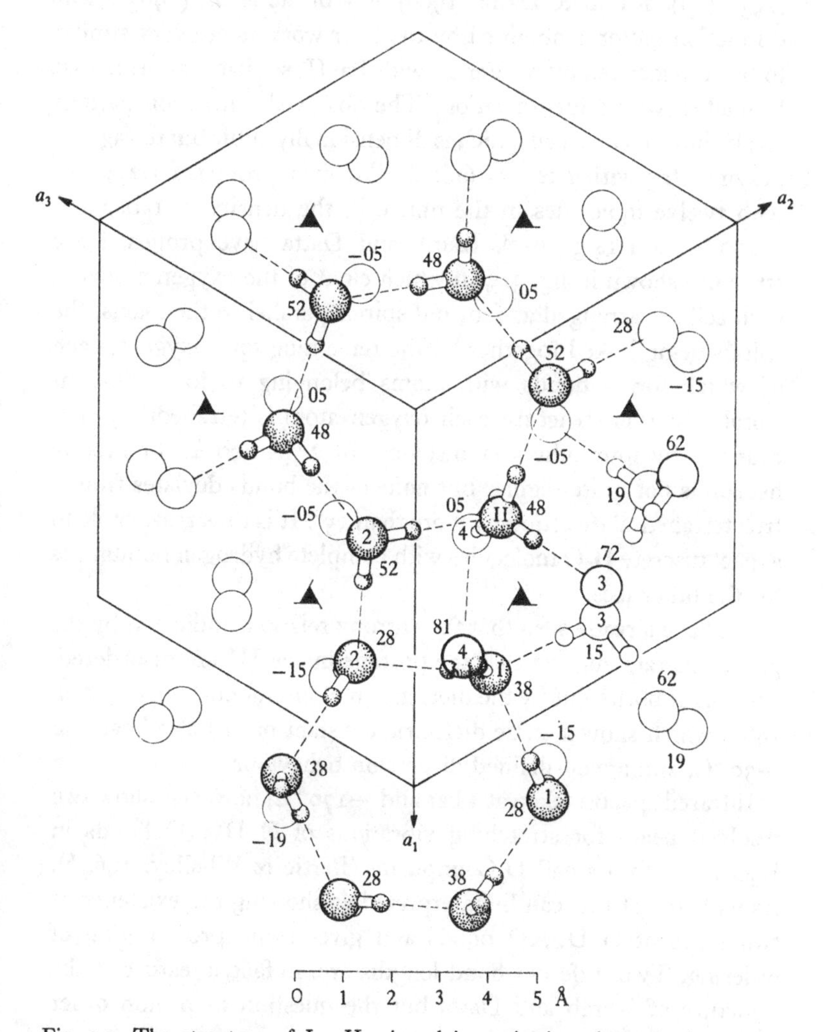

Fig. 3.3. The structure of Ice II, viewed in projection along the hexagonal *c*-axis. The rhombohedral unit cell is outlined and threefold screw axes are indicated. Figures give the height of each atom above the projection plane in units of hundredths of the *c*-axis length. Some water molecules are shown with protons, in the ordered arrangement deduced crystallographically; the hydrogen bonds are shown by broken lines. Non-equivalent water molecules are labelled I and II. (Kamb, 1964.)

3.5. Ice III and Ice IX

Diffraction data on Ice III have been obtained by McFarlan (1936*a*, *b*), Kamb & Datta (1960) and Bertie *et al.* (1963). The diffraction patterns obtained by the later workers are very similar to those associated by McFarlan with Ice II, so that it appears that he made a wrong identification. The observed diffraction pattern can be indexed on a cell which is dimensionally cubic but tetragonal in symmetry, with $a \simeq c = 6{\cdot}80$ Å. The space group is $P4_12_12$ and, with twelve molecules in the unit cell, the density at 1 bar and -190 °C is $1{\cdot}14$ g cm^{-3}. Kamb and Datta have proposed the structure shown in fig. 3.4, in which eight of the oxygen atoms in each cell are arranged as fourfold spirals parallel to the *c*-axis, the spirals being linked together by the remaining four oxygens, each of which forms bonds with atoms belonging to four different spirals. In this structure each oxygen atom is tetrahedrally surrounded by four others at distances of 2·73–2·90 Å. The tetrahedron is not quite regular but none of the bonds deviates from a true tetrahedral direction by more than 15°. It is thus reasonable to expect discrete H_2O molecules with complete hydrogen bonding as for the other ices.

We have already seen that the entropy relations indicated by the phase diagram suggest that the protons in Ice III are disordered, and this is borne out by the dielectric measurements (Wilson *et al.* 1965) which show a static dielectric constant of 117 at 3 kbar and -30 °C, and a well-defined dispersion behaviour.

Infrared spectra taken at 1 bar and -170 °C, however, show two resolved peaks for stretching vibrations of O–D...O bonds in H_2O ice with a small D_2O impurity (Bertie & Whalley, 1964*b*). As with Ice II this can be interpreted as showing the existence of two types of O–D...O bonds and gives some presumption of ordering. Two different bond lengths are, in fact, a feature of the structure of Kamb and Datta but the question of proton order requires further examination.

Since it is known that Ice I_h has proton disorder at all temperatures, a study of the I–III transition entropy as a function of temperature should give information about the amount of ordering in Ice III. There will certainly be some contribution to this entropy change from the difference in lattice vibrational spectrum between the two forms but this should be small compared with any ordering entropy. The I–III transition can, in fact, be fol-

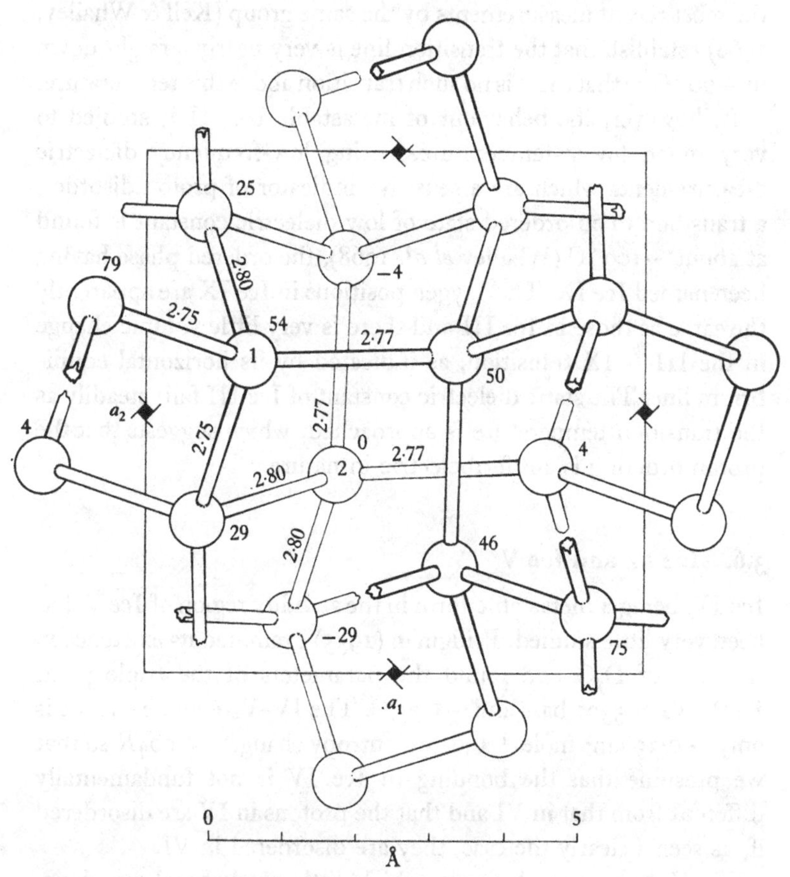

Fig. 3.4. Oxygen positions in Ice III and its proton-ordered analogue Ice IX. The drawing is a projection along the c-axis with heights above the projection plane given in hundredths of the c-axis (c = 6·83 Å). (Kamb, 1968.) (From *Structural Chemistry and Molecular Biology*, ed. Alexander Rich and Norman Davidson. San Francisco: Freeman & Co., copyright © 1968.)

lowed to quite low temperatures because Ice III can be formed metastably in the region of II, as indicated by the dotted line in fig. 3.1. The relation between the slope of this phase boundary and the entropy of transition is given by the Clausius–Clapeyron equation (3.3). Both Tammann and Bridgman and, more recently,

Whalley & Davidson (1965) reported a change in slope of the I–III transition line near −40 °C, suggesting an order–disorder transition among the protons in Ice III near this temperature, but the most recent measurements by the same group (Kell & Whalley, 1968) establish that the transition line is very nearly straight down to −70 °C so that there is no such transition above this temperature.

If, however, the behaviour of metastable Ice III is studied to very much lower temperatures, using low-frequency dielectric measurements which are a sensitive indicator of proton disorder, a transition to an ordered state of low dielectric constant is found at about −100 °C (Whalley *et al.* 1968), the ordered phase having been named Ice IX. The oxygen positions in Ice IX are apparently the same as those in Ice III and there is very little volume change in the III → IX transition, as indicated by its horizontal equilibrium line. The static dielectric constant of Ice III falls steadily as the transition temperature is approached, which suggests that the proton ordering is antiferroelectric in nature.

3.6. Ice IV and Ice V

Ice IV, being a metastable form in the stability region of Ice V, has been very little studied. Bridgman (1935) examined its existence in the case of D_2O and found the parameters of the triple point L–IV–VI at 5300 bars and −6·2 °C. The IV–VI volume change is only −0·35 cm^3 mole^{-1} and the entropy change $-0{\cdot}004R$ so that we presume that the bonding in Ice IV is not fundamentally different from that in VI and that the protons in IV are disordered if, as seems clearly the case, they are disordered in VI.

Ice V, too, is a phase on which little work has been done. X-ray observations have been made (Bertie *et al.* 1963) and Kamb *et al.* (1967) have proposed a monoclinic structure belonging to space group $A2/a$ with a unit cell of dimensions

$$a = 9{\cdot}22\ \text{Å}, \quad b = 7{\cdot}54\ \text{Å}, \quad c = 10{\cdot}35\ \text{Å}, \quad \beta = 109{\cdot}2^\circ, \qquad (3.6)$$

containing twenty-eight molecules arranged as shown in fig. 3.5. The infrared spectrum (Bertie & Whalley, 1964*b*) indicates, as with the other ices, intact H_2O molecules and essentially complete hydrogen bonding, while the lack of resolved peaks on the O–D...O stretching bands indicates either a large range of slightly differing bond characters or else the effects of proton disorder. The small entropy difference between Ices III and V, as shown in table 3.1,

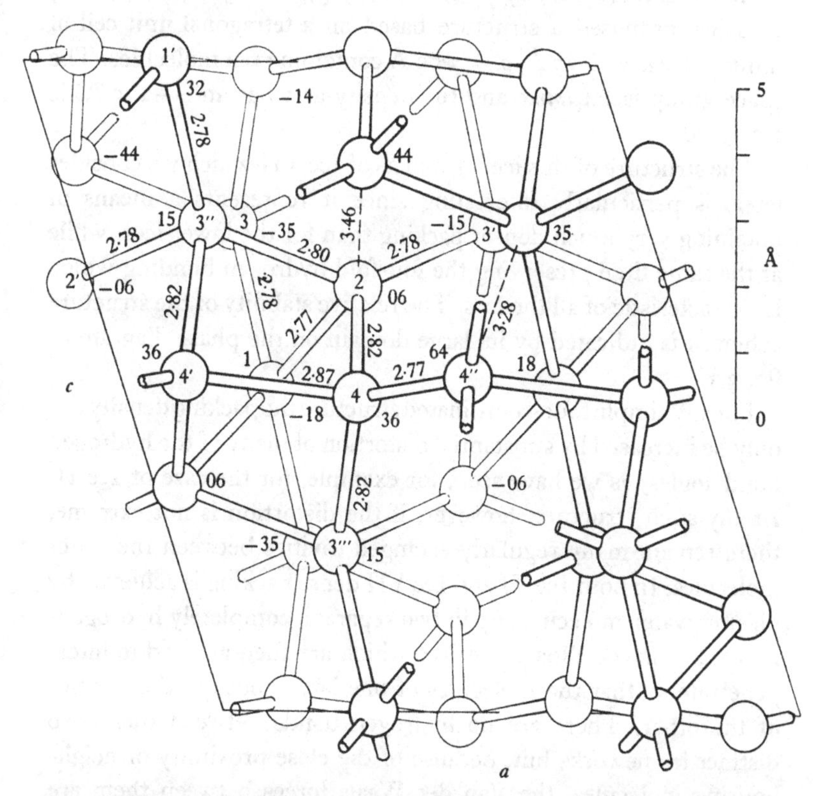

Fig. 3.5. Structure of Ice V projected along the *b*-axis. Heights are given in hundredths of the *b*-axis length ($b = 7{\cdot}54$ Å) and non-equivalent atoms are differently numbered (Kamb, 1968). (From *Structural Chemistry and Molecular Biology*, ed. Alexander Rich and Norman Davidson. San Francisco: Freeman & Co., copyright © 1968.)

and the value $+0{\cdot}58R$ for the transition II → V confirm the existence of proton disorder and final confirmation comes from the dielectric dispersion and high static dielectric constant shown in table 3.3. It is, of course, possible that there is a transition to an ordered state at a temperature below those studied.

3.7. Ice VI

The crystal structure of Ice VI has been studied by X-ray methods by Bertie *et al.* (1964), by Weir *et al.* (1965), and by Kamb (1965*a*) who has proposed a structure based on a tetragonal unit cell of dimensions $a = 6{\cdot}27$ Å, $c = 5{\cdot}79$ Å containing ten molecules. The space group is $P4_2/mmc$ and the density at 1 bar and −175 °C is 1·31 g cm^{-3}.

The structure of this ice, as well as of Ice VII which we consider later, is particularly interesting since it represents a means of obtaining very much denser packing than for the lower ices, while at the same time preserving the fourfold hydrogen bonding which is characteristic of all the ices. The relative stability of the structure achieved is indicated by its large domain on the phase diagram of fig. 3.1.

For any simple four-coordinated structure the packing density can only be increased by substantial distortion of many of the hydrogen bond angles, as we have seen, for example, for the case of Ice II. In any such structure, however, if the distortion is not extreme, there remain many regularly arranged cavities between the water molecules. In both Ice VI and Ice VII dense packing is achieved by placing water molecules upon two separate, completely hydrogen-bonded four-coordinated lattices which are then allowed to interpenetrate so that the molecules of one lattice occupy the cavities of the other. There are no hydrogen bonds between these two distinct frameworks but, because of the close proximity of neighbouring molecules, the Van der Waals forces between them are quite high (Kamb, 1965*b*). In the case of Ice VI the individual frameworks can be thought of as hydrogen-bonded chains of water molecules running parallel to the *c*-axis and bonded transversely to similar chains arranged in a square pattern as shown in fig. 3.6. The second framework, one chain of which is seen in the centre of the figure, is identical with the first and interlaces with it. There are two sets of non-equivalent oxygen atoms in each framework and there is considerable bond distortion, angles ranging from 76° to 128°, but the great increase in packing density more than compensates for this.

The Ice VI structure can be described as a 'self-clathrate', by analogy with the clathrate compounds which are the hydrates of simple substances like noble gases or small hydrocarbon molecules (Pauling, 1960, pp. 469–72). An example of one of these is chlorine

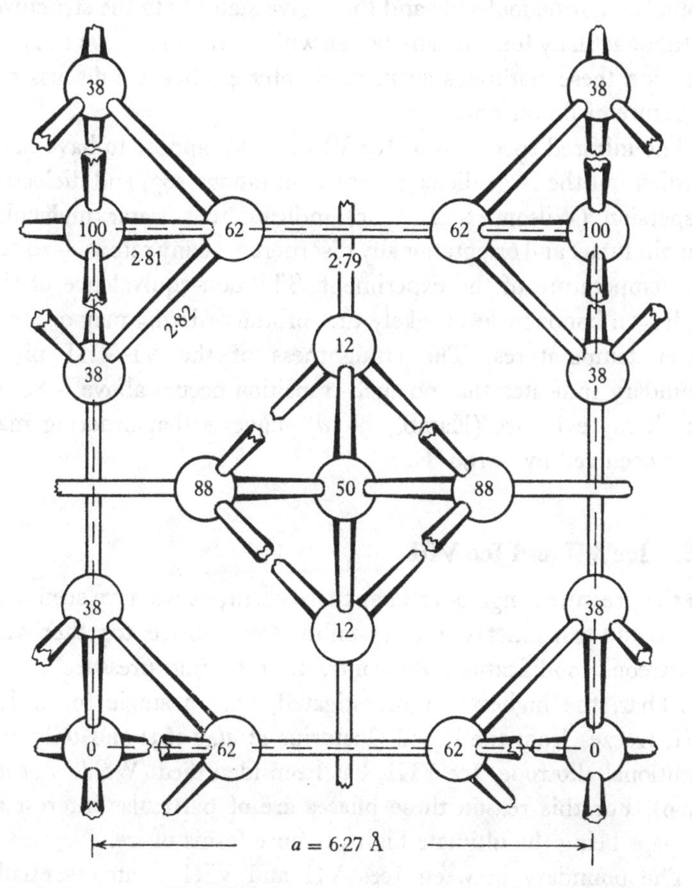

Fig. 3.6. Structure of Ice VI projected along the *c*-axis with heights indicated as before (c = 5·79 Å). Molecules near the cell corners are linked together in chains forming one framework and part of the second framework is seen at the centre of the cell (Kamb, 1968). (From *Structural Chemistry and Molecular Biology*, ed. Alexander Rich and Norman Davidson. San Francisco: Freeman & Co., copyright © 1968.)

hydrate, which is based on units of twenty water molecules bonded together to form a pentagonal dodecahedron, this involving relatively little distortion of the tetrahedral bond angles. Two such dodecahedra together with six other water molecules are bonded together to form a cubic structure, the spaces between the dodecahedra having the form of polyhedra with two hexagonal and eight pentagonal faces. Each of these large figures can accommodate a single chlorine molecule and these give stability to the structure, without actually forming any bonds with it. We shall have cause to mention these clathrates again in chapter 4 when we discuss the structure of liquid water.

The infrared spectrum of Ice VI does not appear to have been studied but the large dielectric constant (about 193) and dielectric dispersion (Wilson *et al.* 1965) indicate that water molecules remain intact and orientationally disordered, at any rate at −30 °C, the temperature of the experiment. The non-equivalence of the hydrogen bonds makes it likely that proton ordering may occur at lower temperatures. The straightness of the VI–VIII phase boundary indicates that no such transition occurs above −80 °C but X-ray evidence (Kamb, 1965*a*) suggests that ordering may have occurred by −196 °C.

3.8. Ice VII and Ice VIII

In the pressure range 0–22 kbar there occur, as we have seen, five stable and three metastable crystalline forms of ice, together with a vitreous modification. By contrast, from this pressure up to 200 kbar, the highest yet investigated, only a single form, Ice VII, freezes from the liquid (Pistorius *et al.* 1963) and only one additional allotrope, Ice VIII, has been identified (Whalley *et al.* 1966). For this reason these phases are of particular interest as perhaps being the ultimate high-pressure forms of ice.

The boundary between Ices VII and VIII is an essentially horizontal line out to at least 50 kbar and the transition is characterized by zero volume change and an entropy change of $-0{\cdot}47R$ at the VI–VII–VIII triple point. Furthermore, the dielectric behaviour of Ice VII shows dispersion and a static value of about 150 while ice VIII is a non-dispersive simple dielectric with a dielectric constant less than about 5 (Whalley *et al.* 1966). It is therefore reasonable to assume that Ices VII and VIII have the same basic structure but the protons in VII are disordered while those in VIII are ordered.

This conclusion is borne out by the similarity of X-ray diffraction patterns taken in the Ice VII region at 20 °C and 25 kbar (Weir *et al.* 1965) and in the Ice VIII region at −50 °C and 25 kbar (Kamb & Davis, 1964). Both patterns indicate a body-centred cubic cell containing two molecules, the lattice parameter at 25 kbar and 20 °C being $3{\cdot}40 \pm 0{\cdot}05$ Å and, at −50 °C, $3{\cdot}30 \pm 0{\cdot}01$ Å. The density at this pressure and −50 °C is 1·66 g cm^{-3}. The X-ray powder pattern of Ice VII, quenched to liquid nitrogen temperature and relaxed to atmospheric pressure, can be indexed on a similar body-centred cubic cell (Bertie *et al.* 1964) but shows a number of weak extra lines which appear to indicate some sort of superstructure. Ices VII and VIII differ in this from the other ices, which show essentially the same diffraction patterns in the quenched state as in their normal stability domain.

Kamb & Davis (1964) have proposed for Ice VII, and by implication for VIII also, the simple structure shown in fig. 3.7. As in the case of Ice VI, the structure consists of two interpenetrating frameworks of H_2O molecules, each of which is completely hydrogen-bonded within itself and makes no bonds to the other. In the present case each framework is a diamond cubic Ice I_c structure with exact tetrahedral bonding and the interpenetration would give just twice the density of Ice I_c, were it not for the fact that there is repulsive contact between oxygen atoms in the alternative frameworks. This increases the O–H. . .O distance to 2·86 Å in Ice VII at 25 kbar compared with the distance 2·76 Å in Ice I at 1 bar. The mode of proton ordering in Ice VIII is not yet known.

It is interesting to consider, as have Kamb & Davis (1964) and Kamb (1965*b*), the possible existence of an ice polymorph still more closely packed than Ice VII. Some ultimate collapse with destruction of the identity of the molecules must occur at extreme pressures, but we mean here, rather, a close-packed form which maintains some of the physical properties of ice through preservation of molecular identity and hydrogen bonding. To be more closely packed than Ice VII such a structure would need to be a spherical close-packing in either a hexagonal close-packed or a face-centred cubic arrangement. The stability of these structures relative to Ice VII depends on the balance between the volume energy gained by denser packing and the energy penalty involved in hydrogen bond distortion. For hexagonal packing no proper system of hydrogen bond orientations can be built up, but such a system is possible for cubic packing if H–O–H angles of 90° are allowed.

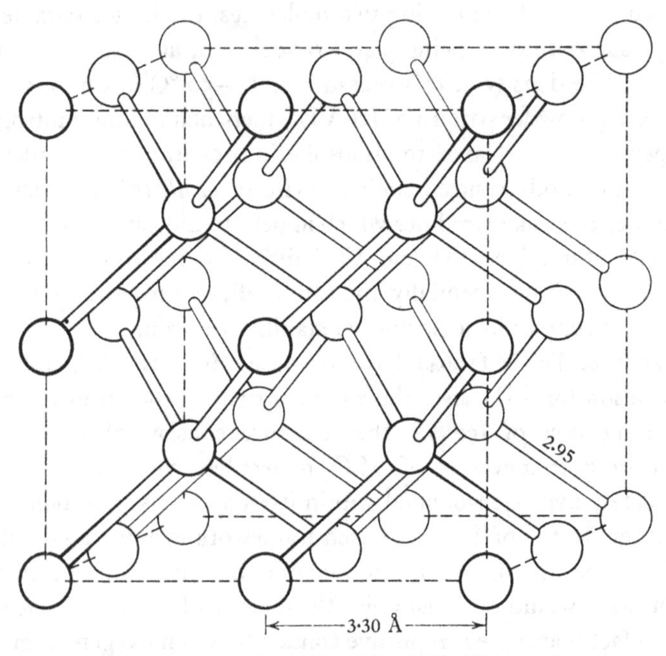

Fig. 3.7. Oxygen positions in Ice VII and Ice VIII. The two interpenetrating Ice I_c frameworks can be seen (Kamb, 1968). (From *Structural Chemistry and Molecular Biology*, ed. Alexander Rich and Norman Davidson. San Francisco: Freeman & Co., copyright © 1968.)

This represents, however, a relatively large bond strain for a rather small gain in packing efficiency. Calculations using reasonable potential functions (Kamb, 1965*b*) confirm that Ice VII is more stable than such a hypothetical close-packed structure for pressures up to at least 200 kbar along the liquid–solid phase boundary.

CHAPTER 4

LIQUID WATER AND FREEZING

Our understanding of the structure and properties of liquids is in a much less well developed state than is the theory of solids and, for that reason if no other, it is difficult to give a concise view of the subject. Simple liquids like argon or sodium, which behave to a first approximation like an assembly of hard spherical atoms, can be treated reasonably satisfactorily by current methods but the non-spherical nature of the water molecule leads to molecular association in the liquid state which complicates the problem immensely.

In this book, which is primarily about ice, we shall be concerned with only a few aspects of the structure and behaviour of liquid water; a comprehensive discussion of water and aqueous solutions would occupy several volumes. In particular we shall discuss current views on the structure of water at temperatures not too far removed from the normal freezing point and then go on to consider in some detail the phase transition involved in freezing. The actual kinetics of crystal growth will be reserved for discussion in chapter 5. Among reviews of the liquid state which provide useful background are those of Green (1960), Furukawa (1962), Barker (1963), Kavanau (1964) and Pryde (1966).

4.1. Experimental information on water structure

In the case of a crystalline solid it is possible to determine, by diffraction methods, the equilibrium positions and vibrational amplitudes of all the atoms involved and this information specifies the structure of the crystal. A liquid, by its very nature, cannot have a structure in this sense. The environment of each atom or molecule is continually changing and we must usually be satisfied with some sort of time-averaged specification of the environment or, which is essentially the same thing in this case, a space average over the environments of many different molecules.

Restricting ourselves for the moment to simple monatomic

liquids, this average information can be expressed in terms of a set of correlation functions $g_n(\mathbf{r}_1; \mathbf{r}_2, \mathbf{r}_3, \ldots, \mathbf{r}_n)$ which specify the probability that an atom be found at the position $\mathbf{r}_1$, given that there are atoms at positions $\mathbf{r}_2, \mathbf{r}_3, \ldots, \mathbf{r}_n$. For a molecular liquid such as water these functions g_n can be generalized by adding to each position specification $\mathbf{r}_i$ an index $\boldsymbol{\alpha}_i$ specifying the orientation of the molecule at this position. The most important of the functions g_n, and the only one which can be found directly from experiment, is the two-body correlation function $g_2(\mathbf{r}_1; \mathbf{r}_2)$. This is usually simplified by taking $\mathbf{r}_2 = 0$ and making use of the fact that $g_2(\mathbf{r}_1; 0)$ is spherically symmetric to write

$$n_L g(r) = \frac{1}{4\pi} \iint g_2(\mathbf{r}; 0)\, d\theta\, d\phi, \tag{4.1}$$

where n_L is the number density of atoms in the bulk liquid and $g(r)$ is called the radial distribution function. The probability of finding an atom at a distance between r and $r+dr$ from a given atom is thus

$$n(r)\, dr = 4\pi r^2 n_L g(r)\, dr. \tag{4.2}$$

This very incomplete statistical information is all that can be found out directly about liquid structure. For a simple liquid $g(r)$ is zero below some small value of r, the hard-sphere radius, rises sharply to a peak showing the existence of a nearest-neighbour shell and then, after more minor oscillations, tends to a steady value of unity. The nearest-neighbour shell usually contains about eleven molecules, in comparison with the twelve characteristic of a close-packed solid, and this accounts for the lower density of the liquid and allows for its fluid properties.

The X-ray diffraction pattern of liquid water has been determined by several workers (Morgan & Warren, 1938; Brady & Romanow, 1960; Danford & Levy, 1962), who agree upon the radial distribution curves shown in fig. 4.1. Since protons are not seen, these curves give the distribution of oxygen atoms. The first maximum, representing the nearest-neighbour shell, is at 2·90 Å at 1·5 °C and this distance increases to 3·05 Å at 83 °C, in comparison with the nearest-neighbour distance of 2·76 Å in ice. The co-ordination in water is, however, rather higher than in ice, the average number of molecules in the nearest-neighbour shell ranging from about 4·4 at 1·5 °C to 4·9 at 83 °C. This more than compensates for the greater nearest-neighbour separation and gives the liquid a density about 8 per cent greater than that of the solid. It is immediately clear from these figures that water is very far from being a simple, close-packed liquid and indeed they suggest that there must be a good deal of four-coordinated structure left.

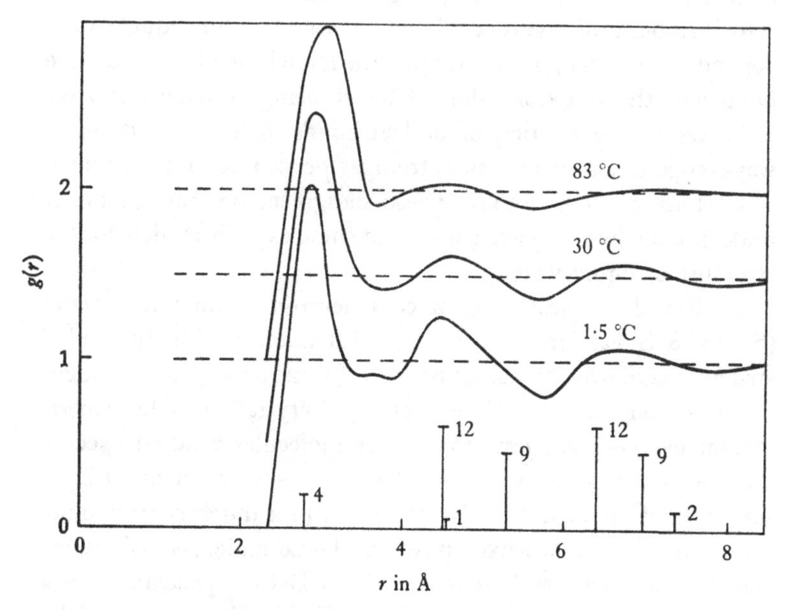

Fig. 4.1. The radial distribution function $g(r)$ for liquid water at several temperatures. The origins of successive curves are displaced upwards by 0·5 and the broken line is the uniform distribution $g(r) = 1$. Vertical lines at the bottom of the diagram give the numbers and positions of neighbours in Ice I_h. (Data from Morgan & Warren, 1938.)

Experimental evidence bearing on this point is, as we shall see, indirect and allows considerable latitude in building up a picture of the instantaneous structure of water.

The other major sources of experimental information about water structure are, as in the case of ice, from dielectric and infrared spectral studies. The dielectric constant of water is high, about 80, which indicates that the molecules are intact and free to rotate. The relaxation peak, above which the dielectric constant falls to a low value, occurs at microwave frequencies ($\lambda \sim 3$ cm at 0 °C, and ~ 0·6 cm at 75 °C; Hasted, 1961), which indicates that, whatever structures may exist in water, they fluctuate quickly enough to allow reorientation of a water molecule in a time of order 10^{-11} s.

Infrared and Raman spectra also confirm the existence of H_2O molecules and, upon more detailed interpretation, can give information about the extent of hydrogen bonding in the liquid. The actual amount of hydrogen bonding assessed for liquid water depends very much, however, upon the model used for a hydrogen bond and the criterion adopted for it being 'broken'. Indeed, estimates of the fraction of broken bonds obtained in different ways range almost continuously from 2·5 per cent to 71·5 per cent at 0 °C (Falk & Ford, 1966). These disagreements can be better understood when we have considered some explicit models for the structure of liquid water.

Studies of the scattering of cold neutrons from liquid water (Singwi & Sjölander, 1960) confirm that water is a highly bonded structure and, over very short periods of time, may even be thought of, from some points of view, as quasi-crystalline. The experimental evidence suggests that water molecules tend to execute vibrations of frequency $\sim 10^{13}$ s^{-1} about positions which are fixed for times of about 4×10^{-11} s and then, in a much shorter time, diffuse to a new quasi-fixed position. These molecular vibrations can be reasonably well described by a Debye spectrum with a characteristic temperature of about 155 °K (Joshi, 1961), while the free diffusional motion corresponds to a rearrangement of the local bonding pattern.

4.2. Structural models

Though the statistical averages expressed by the many-body correlation functions g_n contain a great deal of information, it is much more useful for many purposes to have instead a model giving some sort of instantaneous picture of one particular molecular configuration of the liquid. If this picture is correct and extends over a reasonably large volume, then it contains all the information given by the correlation functions and allows a straightforward evaluation of the partition function and hence of the thermodynamic properties of the liquid. More than this, however, the concreteness of the model leads to a readier interpretation of some of the more subtle properties of the material.

Most of the unusual properties of water derive from the structure of the water molecule which, as we saw in chapter 1, has an approximately tetrahedral shape with protons near two of the vertices and a concentration of negative charge arising from the lone-pair electron hybrids distributed between the remaining two. The interaction of a proton from one molecule with a lone pair from another forms a hydrogen bond—a bond whose properties depend almost as much on the lone-pair electrons as on the proton. In all the crystalline forms of ice the fourfold hydrogen bonding between water molecules plays a major part. The bonds may be straight and closely tetrahedral as in Ices I, VII and VIII or considerably distorted as in the other ices but they appear to maintain their character. It is not surprising, therefore, that they play a major part in determining the structure of liquid water.

There are basically two current approaches to formulation of a model for liquid water. The first, which might be called a homogeneous or uniform model, is exemplified by the treatment of Pople (1951). The characteristic feature of this model is that disorder is introduced into the structure by bending and stretching bonds rather than by breaking them. There is a continuous spectrum of these distortions and all molecules have essentially similar environments. The other group of theories, of which there are many variants, regards water as a mixture of two or more different structures, each of which has a well-defined bonding pattern and exists as clusters of similarly bonded molecules. In both sets of theories, of course, the structure changes very quickly so that the time-average structure round a given molecule is the same as the space average at a given instant.

The fundamental difference between these two approaches arises from a different assessment of the importance of the co-operative aspect of hydrogen bonding. If hydrogen bonds are primarily individual structures involving mainly the electrons concerned in the bond, then individual bonds may be distorted quite arbitrarily, subject only to the distortion energies involved. If, however, there is a large amount of co-operation involved in bonding, either because of the tetrahedral hybridization which it enforces upon bonded molecules or equivalently because of off-diagonal terms in the molecular polarizability tensor, then there will be a tendency to form fully-bonded clusters of water molecules, stabilized by these interactions. The clusters themselves will not be permanent entities but will continuously form and dissociate under the influence of thermal fluctuations. This 'flickering cluster' model is generally associated with the names of Frank & Wen (1957; and also Frank, 1958) who proposed it in some detail, but mixture models of one sort or another have a quite long history.

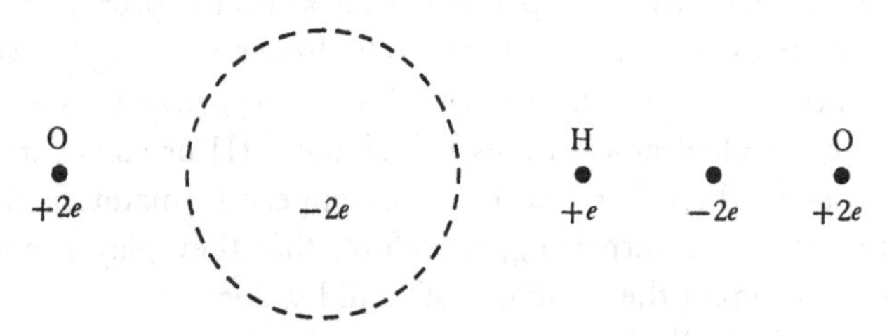

Fig. 4.2. The electrostatic model of the hydrogen bond O–H...O used by Pople (1951) to calculate bond-bending energy.

Apart from the general question of co-operative bonding, the uniform model of Pople (1951) stands or falls on the energy required to bend the hydrogen bond joining two molecules which are already participating in hydrogen bonds to other molecules. This is a complex problem, as our brief discussion of the normal hydrogen bond in chapter 2 has shown, and no reliable force constant is yet known. Pople sidestepped the quantum-mechanical problem by using an electrostatic model for the bond of the type shown in fig. 4.2. The lone-pair electrons are distributed over a sphere in rough approximation to a tetrahedral orbital and the total energy of the bond is 0·26 eV. If the energy of the bond as a function of the distortion $\delta\phi$ of the O–H–O angle is approximated by

$$U(\delta\phi) = U_0 - B\cos\delta\phi \tag{4.3}$$

then, at 0 °C, the value of B is estimated to be between $10kT$ and $13kT$. Higher values are obtained if the lone-pair charge distribution is contracted to a point. If the value of $10kT$ for B at 0 °C is accepted, then a simple application of Boltzmann statistics gives average angular distortions, defined as $\cos^{-1}\langle\cos\delta\phi\rangle$, of 26° at 0 °C and 30° at 100 °C. These distortions are so large that any ice-like molecular arrangement cannot extend much past second-nearest neighbours and the structure is acceptably liquid-like if continual rearrangements take place.

Since every model must be consistent with the radial distribution function, this gives further data on which to base the model. The bond lengths must have a distribution determined by the stretching force constant of the bond and this could be calculated. Pople chose rather to adopt an arbitrary but reasonable Gaussian distribution of standard deviation 0·258 Å about a mean bond length of 2·80 Å to best fit the first peak of the radial distribution curve. With this assumption and an adjustment for double counting of more distant neighbours, the model gives very good agreement with the experimental curves out to about 5 Å, both at 1·5 °C and 83 °C. In evaluating the significance of this fit, however, the fact that B and the bond length and its standard deviation at the temperature considered are essentially arbitrary parameters must be borne in mind. It is interesting to note that, whilst the number of hydrogen-bonded nearest neighbours is exactly four, the bond distortions allow some more distant neighbours to penetrate the nearest-neighbour shell, giving a co-ordination number in excess of four, as observed. This structural collapse also explains why the disordered structure, representing liquid water, has a greater density than crystalline ice. The consequences of Pople's model do not appear to have been worked out in much more detail than this so that it is not known how well it predicts other thermodynamic quantities.

Apart from the general indication from studies of the high-pressure ices that bond-bending is favoured over bond-breaking, some of the strongest evidence in support of such a uniform model comes from the Raman and infrared studies of Wall & Hornig (1965) and of Falk & Ford (1966) respectively. By using a dilute solution of D_2O in H_2O they were able to examine essentially uncoupled O–D bands and hence to probe the average deuteron environment in the liquid. The infrared bands showed a symmetrical Gaussian contour for each of the three fundamentals, rather than the split or asymmetric bands which would be expected if there were a sharp distinction between bonded and unbonded molecules. Similar simple behaviour was shown by the Raman bands. Infrared and Raman evidence by other workers which supports the alternative bond-breaking models is generally much less clear-cut, though recent work by Walrafen (1968) using D_2O in H_2O does show an asymmetry in the O–D stretching band at 2500–2600 cm^{-1} which invites interpretation in terms of two distinct bonding states.

Another type of uniform model which concentrates attention upon the nearly tetrahedral environment of each water molecule but permits bonds to simply break rather than bend has been discussed by Orentlicher & Vogelhut (1966) and leads to quite good agreement with experiment for several thermodynamic properties. Sparnaay (1966) too has considered some of the properties of an intermediate model in which the cosine variation (4.3) for bond energy is replaced by a step function, so that after a limited angular deflexion the bonds break. Once again, however, the consequences of this model have not yet been fully explored.

Turning now to the mixture models, we find a great variety, ranging from early views of water as a simple mixture of polymers $(H_2O)_n$ to highly developed and highly structured groupings. One of the earliest and best known of these models is that of Bernal & Fowler (1933), who proposed that water consists of molecular groupings of three different kinds: (i) an open structure like Ice I or the silica mineral tridymite, this structure being rather rare but present to some extent below 4 °C, (ii) a more dense quartz-like structure predominating at ordinary temperatures, and (iii) a close-packed ideal ammonia-like liquid of most importance at high temperatures approaching the critical point. This mixture is able to reproduce, to a reasonable degree, the radial distribution function and gives a natural explanation of other aspects of the behaviour of water. For example, the density maximum at 4 °C arises because, as the temperature rises from 0 °C, the ice-like structure is being broken up in favour of the more dense, quartz-like arrangement whose natural thermal expansion dominates at higher temperatures. Bernal and Fowler did not explicitly describe the water structure in terms of molecular clusters of these three different bonding types but this is implied by the very recognition of distinct bond geometries.

More recent treatments have used somewhat similar models, though with different significant structures in the mixture and with more explicit assumptions about cluster sizes. By setting up a partition function for the system it has then been possible to derive thermodynamic functions for comparison with experiment. One of the best known of these treatments is that of Némethy & Scheraga (1962), who took a model consisting of compact clusters of tetrahedrally bonded molecules, thus having a structure more or less like that of Ice I, together with a fraction of unbonded molecules. On the surfaces of the clusters there are, of course, molecules making two or three bonds, but isolated dimers, trimers and similar small groupings are ruled out. The partition function was evaluated by assigning a constant binding energy to each hydrogen bond and supplementing this with information derived from the known partition functions for ice and for restricted rotation of unbonded water molecules. The energy involved in breaking a hydrogen bond in this sense is clearly less than the hydrogen-bond energy calculated in chapter 2, since the Van der Waals interactions are not destroyed and may even be increased by higher co-ordination. The bond-breaking energy giving best agreement with experiment was found to be 0·057 eV which, while only about one-fifth of the total hydrogen-bond energy, is perhaps not unreasonable in view of its definition. With this bond-breaking energy and the free volume available to the unbonded molecules as adjustable parameters, the model is able to give good agreement with experiment for the radial distribution function, density (including the maximum at 4 °C), free energy and entropy as functions of temperature. The error in the specific heat amounts to less than ± 25 per cent over the whole range 0–100 °C.

Since this model is clearly in some sense a good approximation to reality, it is interesting to examine its conclusions about cluster size and fraction of unbroken bonds at various temperatures. These quantities are displayed in fig. 4.3. The cluster size ranges from 90 molecules at 0 °C to 21 at 100 °C and the fraction of unbroken bonds from 53 per cent down to 33 per cent. The fraction of unbonded molecules varies from 24 per cent at 0 °C to 44 per cent at 100 °C so that, over this whole range, the formation of clusters is a dominant feature of water structure.

Support for this model has been drawn from infrared and Raman studies by Cross *et al.* (1937), Buijs & Choppin (1963), Thomas & Scheraga (1965) and Walrafen (1966, 1968) but this rests upon an assignment of certain marginally resolvable bands to molecular species with different degrees of bonding and is contradicted by the measurements on water/heavy-water mixtures by Wall &

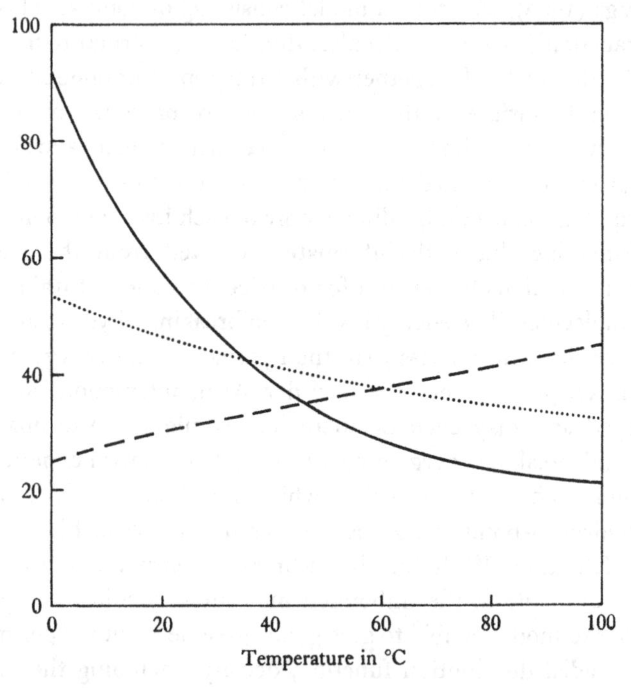

Fig. 4.3. Average cluster size, percentage of unbroken bonds and percentage of unbonded molecules in liquid water as calculated by Némethy & Scheraga (1962); —, Cluster size; – – –, % unbonded molecules; . . . . . ., % unbroken bonds.

Hornig (1965) and by Falk & Ford (1966) to which we have already referred. As we shall see later, however, this conflict can be resolved in part by some reinterpretation.

A step in this direction, which involves a relaxation of the assumption that all hydrogen bonds are equivalent and either fully formed or completely broken, was made by Vand & Senior (1965). Instead of using five sharp molecular levels corresponding to bonding states from 0 to 4 as did Némethy and Scheraga, this model considers only three bonding states, 0, 1, 2, referring to the two protons on a given molecule, but allows the energy of these bonds to have a Gaussian distribution because of stretching or deformation. This model—at any rate for the quantities calculated, including the specific heat—gives results significantly closer to experiment than those of the Némethy–Scheraga model. It is also clearly possible to include two further energy bands specifying the bonding of the lone pairs if desired, though at the present stage this would introduce at least one extra undetermined coefficient.

Similar calculation techniques have been used with other choices of the significant structures involved, the different models giving varying but usually good agreement with experiment. Thus Davis & Litovitz (1965) used a model based on six-membered rings as in ice, the two possible structures being an ice-like arrangement of two parallel rings, giving maximum hydrogen bonding, and a close-packed arrangement achieved by rotating one ring through $\frac{1}{3}\pi$ and then bringing the two closer together. The calculated 'ice-likeness' of water on this basis is 60 per cent at 0 °C and 30 per cent at 100 °C, the fraction of unbroken bonds at 0 °C being 82 per cent. Other models using slightly different mathematical techniques for their evaluation have been proposed by Marchi & Eyring (1964), the significant structures being tetrahedrally bonded ice-like clusters and freely rotating monomers, and by Jhon *et al.* (1966), whose model involves Ice I-like clusters of about forty-six molecules dispersed in a denser Ice III-like medium. In this later model, which gives particularly good agreement with experiment, only the number of clusters is allowed to vary, their size being constant and roughly equal to the average cluster size in the Némethy–Scheraga treatment.

It is perhaps worth while to have a further look at the ice-like clusters which play such an important part in all these theories. In the interests of simplicity they have generally been considered to have a crystalline structure like that of Ice I_h or I_c but such a requirement is unnecessary and, in view of the supercooling of liquid water which we shall consider presently, unlikely. From the point of view of the cluster theories discussed above, the only requirement on the clusters is that they have substantially tetrahedral bonding and the same sort of molecular volume and vibrational spectrum as ice clusters of similar size. Pauling (1959) has suggested that cages of water molecules, such as those formed in clathrate compounds around molecules like chlorine, may form particularly stable and therefore preferred bonding arrangements in liquid water. Such clathrate cages consist of twenty molecules lying at the corners of a pentagonal dodecahedron and forming a total of thirty hydrogen bonds. An additional molecule, making no bonds at all, can be accommodated in the centre of the cage. Larger clathrate complexes containing forty-one molecules are also possible. It is perhaps significant that these structures are of about the size required by most of the cluster theories so that, whilst it would probably be too extreme to build a model solely upon them, they may well form a considerable fraction of the molecular associations in the liquid. Models emphasizing the importance of these structures have been discussed by Frank & Quist (1961) and by Malenkov (1961).

From these diverse models of the structure of liquid water it is now perhaps possible to make a synthesis which, while rather qualitative, is not too far from reality. Hydrogen bonding and near-tetrahedral co-ordination are certainly dominant features of the water structure and, whether the arrangement of molecules is thought of in terms of sharply defined clusters or in terms of continuously distorted hydrogen bonds, the average coherence length is several molecular diameters and the number of molecules involved in this short-range order is several tens. It is also agreed that the cluster size, or the correlation distance, decreases as the temperature is raised and the liquid becomes more closely packed. It seems certain that neither a strict bond-breaking model nor a model which ignores co-operative effects can be exactly true—the cluster size variations of Némethy and Scheraga and the bond bending of Pople's model both measure the same physical effect in different approximations. The clusters actually existing in liquid water are certainly less clearly defined than those used for the sake of formal simplicity in the models while, from the opposite viewpoint, any liquid model which involves no significant fraction of essentially broken bonds is topologically very restrictive unless it adopts an extreme definition of what constitutes a bond.

The clusters or distorted bonding structures are also certainly transient and have short mean lifetimes. If the clusters are to have any meaningful existence, these lifetimes must be long in comparison with molecular vibration periods ($\sim 10^{-13}$ s), but at the same time they must be short enough to account for the relaxation times for dielectric absorption, nuclear spin-lattice relaxation and cold neutron scattering, all of which are $\sim 10^{-11}$ s. Lifetimes of order 10^{-11} to 10^{-10} s as suggested by Frank & Wen (1957) seem very reasonable for either type of model.

4.3. Homogeneous nucleation of freezing

When a small quantity of a liquid is cooled below its equilibrium temperature it does not immediately freeze but remains for some time in a metastable supercooled state. Except in the case of materials which form glasses, it is not possible to maintain the supercooled state indefinitely or to achieve more than a limited degree of supercooling before spontaneous crystallization occurs. It is found experimentally that the maximum supercooling attainable increases with increasing purity of the liquid, particularly as far as foreign particles in suspension are concerned, and that it is easier to supercool small droplets than larger volumes of liquid. Both these effects are easily accounted for by the assumption that foreign solid particles assist the freezing transition, division of a sample of liquid into many small droplets effectively isolating the most active foreign particles into a small fraction of the total number of droplets. Despite the most rigid purification procedures and the use of very small droplets, the amount of supercooling is still found to be limited and most liquids under such conditions freeze spontaneously at about 0·8 T_F, where T_F is their equilibrium absolute freezing temperature (Ubbelohde, 1965). This limit therefore presumably represents some intrinsic property of the liquid.

This behaviour can be very simply understood from consideration of the process of freezing. When a liquid is supercooled it is energetically favourable for it to change to the crystalline state. This cannot be accomplished discontinuously; first a very small volume of liquid must crystallize and this crystal must then grow until all the liquid has frozen. A small crystal embryo is, however, in an energetically unfavourable state because of its very large surface-to-volume ratio and the positive free energy associated with its interface with the liquid. There is thus a free energy barrier to be overcome before freezing can commence and this barrier can only be surmounted by a nucleation process depending upon thermal fluctuations. The stability of a small crystal embryo may be enhanced if it grows closely upon an insoluble foreign particle and in this case the nucleation process is termed heterogeneous. If there are no foreign particles or surfaces present, then freezing must commence by a process of homogeneous nucleation within the pure liquid itself.

Experimental studies of the freezing of water droplets have been particularly extensive because of the importance of this process in the initiation of rainfall from clouds in continental areas (Fletcher, 1962*a*). Such clouds typically consist of droplets 20 μ or so in diameter in concentrations of several hundred per cubic centimetre. For aerodynamic reasons it is very difficult for droplets of this size to collide and coalesce to form raindrops and the cloud may grow to heights well above the freezing level. When the cloud temperature falls to near -20 °C a few of the droplets may freeze and the resulting ice crystals then grow rapidly because of the difference in vapour pressure over ice and water at this temperature. These crystals fall through the cloud, collecting more water droplets by collision, and emerge below the freezing level as raindrops.

In laboratory studies of the freezing process it is necessary to begin with carefully purified water and to preserve it from the influence of its supporting substrate or from airborne contamination during the measurements. Many workers have examined the freezing of water droplets resting on metal or glass plates coated with a hydrophobic film, while recent work has used a method developed by Bigg (1953) in which the droplets are supported at the interface between two immiscible oily liquids. The results of the large number of such experiments have been discussed by Langham & Mason (1958) and fig. 4.4 is an extended version of their plot of observed maximum supercooling as a function of liquid volume. It can be seen that, despite the wide diversity of the results, there is a reasonably consistent lower limit to the temperature which can be reached before nucleation occurs, and it seems reasonable to explain the results of those experiments which gave higher freezing temperatures as due to some form of contamination in the sample. These, then, are the experimental results which, together with more explicit experiments on the

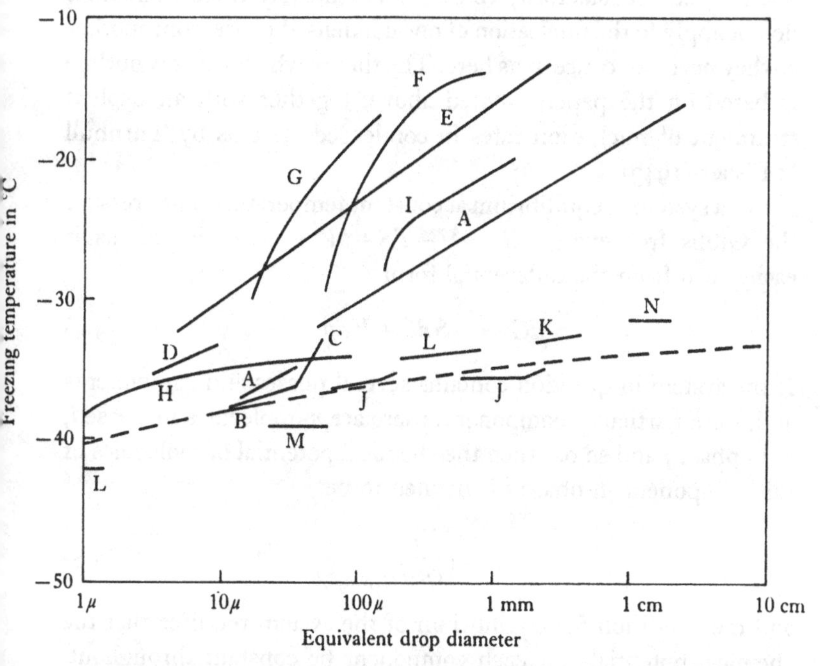

Fig. 4.4. Supercooling obtainable before freezing for pure water droplets as a function of size, as observed by various workers: Bigg (A), Carte (B), Chahall and Miller (C), Day (D), Dorsch and Hacker (E), Heverley (F), Hosler (G), Jacobi (H), Kiryukhin and Pevsner (I), Langham and Mason (J), Meyer and Pfaff (K), Mossop (L), Pound (M), and Wylie (N). The broken curve shows the homogeneous nucleation threshold calculated from theory with $\sigma = 22$ erg cm^{-2}.

effects of foreign nuclei, must be explained by the detailed theory of homogeneous and heterogeneous nucleation.

The foundations of the theory of phase transitions were laid by the work of J. Willard Gibbs towards the end of last century but the detailed application of these ideas to nucleation phenomena was first made by Volmer & Weber (1926; also Volmer, 1939) and then extended by Becker & Döring (1935), Frenkel (1939, 1946) and others; for recent reviews see Hollomon & Turnbull (1953), Turnbull (1956) and Hirth & Pound (1963). Most of this work concerns the condensation of vapours and gives a reasonably satisfactory account of the observed phenomena, though some recent theoretical problems have arisen (Lothe & Pound, 1962; Oriani & Sundquist, 1963; Reiss & Katz, 1967; Abraham & Pound, 1968) which have not yet been satisfactorily resolved. Fortunately these difficulties do not apply to the nucleation of one condensed phase from another so they need not concern us here. The theory which we now outline is based on the papers quoted above, together with an explicit treatment of nucleation rates in condensed systems by Turnbull & Fisher (1949).

For a system in equilibrium at constant temperature and pressure the Gibbs free energy $G = U - TS + pV$ is a minimum, as is easily seen from the differential form

$$dG = -S\,dT + V\,dp. \tag{4.4}$$

If the system in question contains several phases and components and, for a particular component, there are n_i molecules in phase i, n_j in phase j and so on, then the chemical potential of molecules of this component in phase i is defined to be

$$\mu_i = \left(\frac{\partial G}{\partial n_i}\right)_{T,p,n_j} \tag{4.5}$$

and the condition for equilibrium of the system requires that the chemical potentials for each component be constant throughout.

In a liquid at any temperature there will be continual fluctuations in density and structure due to thermal motion. Near the equilibrium freezing point there will be, in addition, heterophase fluctuations resulting in the momentary formation of small groups of molecules in the arrangement characteristic of the solid crystalline phase. In liquid water, as we have seen, there may be a wide variety of hydrogen-bonded clusters differing in size and in molecular arrangement. Among them, those clusters which have an Ice I-like arrangement of molecules are of particular importance since, at temperatures below the equilibrium freezing point, they may occasionally grow by statistical fluctuations to a size at which they are stable relative to liquid water. They can then develop into macroscopic ice crystals.

Consider, in supercooled water, the formation of such an ice-like cluster of i molecules. Since the structure can be extended continuously to bulk ice, the molecules in it can be described by a chemical potential μ_S characteristic of bulk ice at this temperature, together with a correction term to take account of the free energy of the ice/water interface. In the formation of such a cluster, then, the total free energy of the system is changed by an amount

$$\Delta G_i = (\mu_S - \mu_L)i + Ai^{\frac{2}{3}}, \tag{4.6}$$

where μ_L is the chemical potential of molecules in the liquid and the constant A depends upon the shape of the cluster and the structure of the interface. Because the ice structure can be added to without internal bond rearrangement, we can use this equation for ice-like clusters of all sizes and the variation of ΔG_i with i is shown as curve (a) in fig. 4.5. Small clusters are unstable and tend to disappear rather than grow. Once a critical size i^* is reached, however, the addition of further molecules lowers the total free energy of the system and the cluster can develop quickly into a macroscopic crystal. The maximum in the value of ΔG_i, which we call ΔG^*, gives the height of the thermodynamic barrier to nucleation at the particular temperature considered. It is the kinetics of the process of surmounting this barrier which is treated by nucleation theory.

In water there are many possible hydrogen-bonded structures other than those like Ice I and indeed truly ice-like clusters must be comparatively rare to account for the observed metastability of supercooled water. Some clusters in water may correspond to bulk structures having a higher free energy than liquid water and hence are described by a curve like (b) in fig. 4.5. Others, like the clathrate cages of Pauling, have very favourable bonding for particular numbers of molecules but the structure cannot be extended continuously to a bulk phase. Such clusters have a behaviour described by a curve like (c).

For water above the equilibrium freezing point the distribution of clusters can be treated as an equilibrium statistical problem giving the result shown in curve (a) of fig. 4.6,

$$n_i \simeq n_L \exp(-\Delta G_i/kT), \tag{4.7}$$

where n_i is the number of clusters of size i per unit volume and, because of the small concentration of truly ice-like clusters, n_L is essentially the number of water molecules per unit volume. If we try to use (4.7) for the case of supercooled water we get a curve like (*b*) in fig. 4.6 in which the number of very large clusters tends

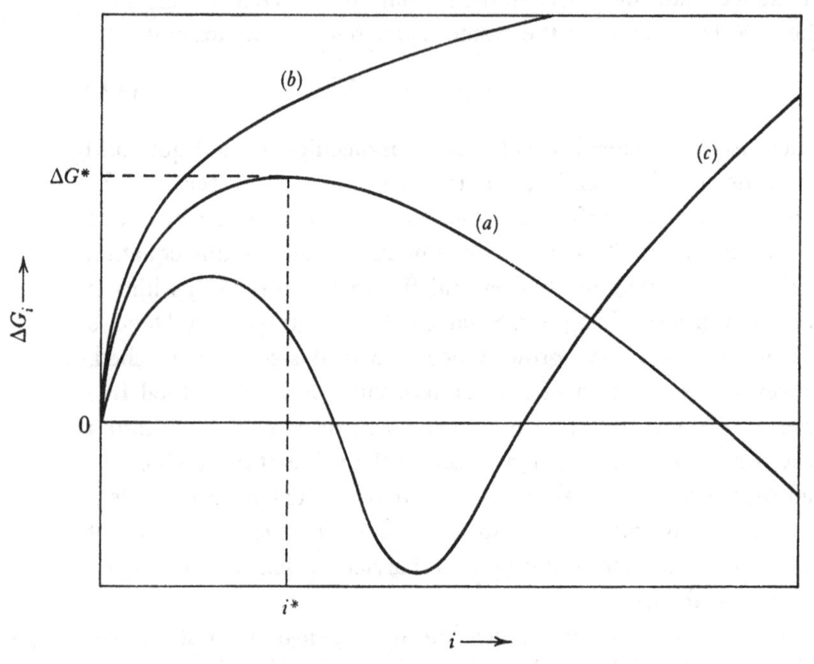

Fig. 4.5. Free energy excess ΔG_i for a cluster containing i molecules (*a*) in an ice-like struture, (*b*) in an energetically unfavourable structure, (*c*) in a structure giving optimum bonding for a small value of i. In all cases the temperature is below 0 °C and the curves have meaning only for integral i.

to infinity. Instead we must therefore use a kinetic approach for this metastable region by considering the growth of a cluster C_i containing i molecules by addition of an unbonded molecule C_1

$$C_i + C_1 \rightarrow C_{i+1} \tag{4.8}$$

and the reverse process

$$C_{i+1} \rightarrow C_i + C_1 \tag{4.9}$$

and maintaining the metastable liquid state by supposing that clusters which grow past a particular size i', which is greater than the critical size i^*, are broken up and returned to the liquid as single molecules. This set of assumptions leads to the stationary distribution of embryos shown as curve (*c*) in fig. 4.6. The net growth current of embryos by reactions (4.8) and (4.9) from C_i to C_{i+1} is independent of i, since the distribution is stationary, and gives the nucleation rate J.

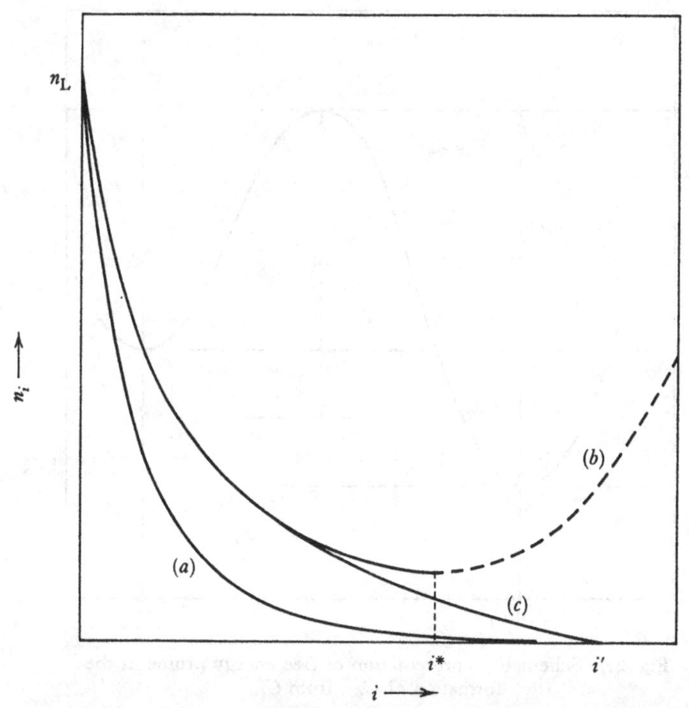

Fig. 4.6. Distribution of ice-like clusters in water (*a*) at a temperature above 0 °C, (*b*) at a temperature below 0 °C if an equilibrium distribution is assumed, (*c*) at a temperature below 0 °C for the steady-state problem.

Turning to the reaction (4.8) as the key to the kinetics of the process, we see that this just involves the addition of one molecule to a cluster. In the case of condensation from the vapour this can be simply evaluated from the probability of collision of a molecule with a small droplet but, for condensed-phase reactions, there are already many molecules essentially in contact with the embryo and it is rather a matter of bonding and reorientation that is involved. Details of the process are essentially the same as those of chemical reaction rates (Eyring, 1935). When a molecule in contact with a cluster C_i is about to join it, the molecule must first pass through an activated state, represented by an unfavourable bond orientation or intermolecular distance, before it joins C_i to form C_{i+1}. The free energy of the system, plotted in terms of this relevant co-ordinate, then has the form shown in fig. 4.7. In terms of any other co-

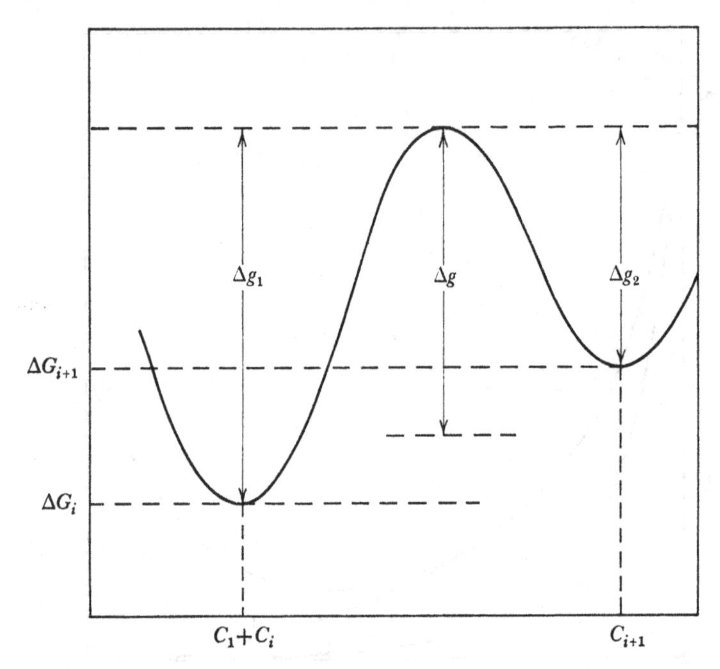

Fig. 4.7. Schematic representation of free energy profile in the formation of C_{i+1} from C_i.

ordinate we find a simple potential well, so that the free energy has a saddle point at the activated complex configuration. This activated state is, in fact, very similar to the intermediate states involved in viscous flow and should have essentially the same activation energy. The rate of the forward reaction is just the probability that one of the contacting molecules is in the activated state multiplied by the velocity with which it moves, as far as the co-ordinate plotted in fig. 4.7 is concerned. For unit volume of liquid this turns out to be just

$$J^+ = n_i \zeta i^{\frac{2}{3}} (kT/h) \exp(-\Delta g_1/kT), \qquad (4.10)$$

where $\zeta i^{\frac{2}{3}}$ is the number of molecules in contact with the surface of C_i, and Planck's constant h arises from the semi-classical evaluation of the partition functions for vibration and translation. Δg_1 is specified in fig. 4.7. In just the same way, the rate for the reverse reaction (4.9) is given by

$$J^- = n_{i+1}\,\zeta(i+1)^{\frac{2}{3}}\,(kT/h)\exp(-\Delta g_2/kT), \tag{4.11}$$

where Δg_2 is also shown in fig. 4.7.

To make progress it is now necessary to suppose that i can be approximated by a continuous variable so that, neglecting the distinction between $i^{\frac{2}{3}}$ and $(i+1)^{\frac{2}{3}}$ in (4.10) and (4.11) and writing

$$\Delta g_1 = \Delta g + \tfrac{1}{2}d(\Delta G_i)/di \tag{4.12}$$

and similarly for Δg_2, where

$$\Delta g = \tfrac{1}{2}(\Delta g_1 + \Delta g_2), \tag{4.13}$$

we find for the net forward rate

$$J = J^+ - J^- = (kT/h)\zeta i^{\frac{2}{3}}\exp(-\Delta g/kT)\left[n_i\exp\left(-\frac{d\Delta G_i/di}{2kT}\right) - n_{i+1}\exp\left(\frac{d\Delta G_i/di}{2kT}\right)\right]. \tag{4.14}$$

If we define a generalized diffusion coefficient D_i by the relation

$$D_i = (kT/h)\,\zeta i^{\frac{2}{3}}\exp(-\Delta g/kT) \tag{4.15}$$

then (4.14) can be written

$$J = -n_i\frac{D_i}{kT}\frac{d\Delta G_i}{di} - D\frac{dn_i}{di}, \tag{4.16}$$

which looks like a generalized mobility–diffusion equation in the space measured by the co-ordinate i, with the driving potential specified by ΔG_i. To solve this equation we follow Frenkel (1946) and rewrite it as

$$J = -D_i\exp(-\Delta G_i/kT)\frac{d}{di}[n_i\exp(\Delta G_i/kT)], \tag{4.17}$$

which can immediately be solved formally for the cluster concentration n_i to give

$$n_i = J\exp(-\Delta G_i/kT)\int_i^{i'} D_i^{-1}\exp(\Delta G_i/kT)\,di, \tag{4.18}$$

where i' is the cluster size at which growing clusters are removed from the system, so that $n_{i'} = 0$.

Now $\exp(\Delta G_i/kT)$ has a sharp maximum at the critical cluster size i^*, so that we may evaluate the integral in (4.18) by replacing D_i^{-1} by $D_{i^*}^{-1}$ and expanding ΔG_i about its value ΔG^* at i^* as

$$\Delta G_i = \Delta G^* + \tfrac{1}{2}\gamma(i-i^*)^2, \tag{4.19}$$

where
$$\gamma = (d^2\Delta G_i/di^2)_{i=i^*}. \tag{4.20}$$

Equation (4.18) can then be inverted to give

$$J = n_i \exp(\Delta G_i/kT) D_{i^*} \exp(-\Delta G^*/kT)(\gamma/2\pi kT)^{\frac{1}{2}}. \tag{4.21}$$

This expression must hold for all i and in particular for very small i where the flux of cluster growth has negligible effect and n_i is given by its equilibrium value (4.7). Inserting this and using (4.15) for D_{i^*} and (4.6) and (4.19) to evaluate γ, we find

$$J \simeq \left(\frac{Aa^2}{9kT}\right)^{\frac{1}{2}} \left(\frac{n_L kT}{h}\right) \exp\left(-\frac{\Delta g}{kT}\right) \exp\left(-\frac{\Delta G^*}{kT}\right), \tag{4.22}$$

which differs only in minor detail from the expression of Turnbull and Fisher (1949). For the freezing of water the factor $(Aa^2/9kT)^{\frac{1}{2}}$ is close to unity and, indeed, for most nucleation problems of interest it lies between 0·1 and 10, so that we can write

$$J \simeq \frac{n_L kT}{h} \exp\left(-\frac{\Delta g}{kT}\right) \exp\left(-\frac{\Delta G^*}{kT}\right), \tag{4.23}$$

where, to recall our original definition, J is the rate of formation of freely growing crystal nuclei in unit volume of liquid.

For the freezing of water $(n_L kT/h) \sim 10^{35}\ \mathrm{cm^{-3}\ sec^{-1}}$. Using for Δg the activation energy for viscous flow of water, which phenomenon continues smoothly into the supercooled region (Hallett, 1963), we find $\Delta g/kT \simeq 10$ at -2 °C and $\Delta g/kT \simeq 15$ at -22 °C and so, by extrapolation, $\Delta g/kT \simeq 18$ in the region of interest from -30 to -40 °C. We can thus rewrite (4.23) for water as

$$J \sim 10^{27} \exp(-\Delta G^*/kT)\ \mathrm{cm^{-3} s^{-1}}. \tag{4.24}$$

As we shall see later, only the order of magnitude rather than the exact value of the pre-exponential factor is of significance.

To find the explicit form of ΔG^* we return to its definition as the maximum value of ΔG_i given by (4.6). The difficulty arises from evaluation of the surface term $Ai^{\frac{2}{3}}$. The coefficient A will certainly depend upon the shape of the embryo and on the topography of its boundary surfaces. It may also depend upon the embryo size, because surface effects generally penetrate at least one or two molecular distances below the geometrical surface. In fact, in a liquid like water, it may be difficult to define clearly the position of the ice-water interface.

These uncertainties are usually sidestepped by ignoring molecular considerations and using a continuum approach, supposing that the embryo has a well-defined size and shape, that the chemical potential μ_S in (4.6) is exactly that of bulk ice and that the interface is characterized by a surface free energy σ identical with that between bulk ice and water. If we define the free energy difference per unit volume between ice and water by

$$\Delta G_v = n_S(\mu_S - \mu_L) \tag{4.25}$$

and suppose the embryo to be a sphere of radius r, then (4.6) becomes

$$\Delta G(r) = \tfrac{4}{3}\pi r^3 \Delta G_v + 4\pi r^2 \sigma \tag{4.26}$$

and its maximum value is easily found to be

$$\Delta G^* = \frac{16\pi\sigma^3}{3(\Delta G_v)^2} \tag{4.27}$$

for a critical radius

$$r^* = -2\sigma/\Delta G_v. \tag{4.28}$$

It is possible, if we wish, to allow σ to be anisotropic, as in a macroscopic crystal, and to envisage embryo shapes other than spherical. The result is that (4.27) is multiplied by a shape factor which is greater than unity and increases as the shape differs increasingly from spherical. In view of the small size of the critical embryo and the other uncertainties involved, such a refinement is not really justified.

To make (4.27) still more explicit we use the thermodynamic relation $S = -(\partial G/\partial T)_p$ to write

$$\Delta G_v = -\int_0^{\Delta T} \Delta S_v \, d\Delta T = -\langle \Delta S_v \rangle \Delta T, \tag{4.29}$$

where $\langle \Delta S_v \rangle$ is an average entropy of fusion over the supercooling range ΔT and, for water, has the approximate value

$$\langle \Delta S_v \rangle \simeq (1{\cdot}13 - 0{\cdot}004\Delta T) \times 10^7 \text{ erg cm}^{-3} \text{ deg}^{-1}. \tag{4.30}$$

We can thus rewrite (4.27) as

$$\Delta G^* = \frac{16\pi\sigma^3}{3(\langle\Delta S_v\rangle\Delta T)^2} \tag{4.31}$$

and, with (4.24), (4.30) and an assumed value of σ, this gives a basis for calculation of the nucleation process.

In the case of condensation of a vapour, where an expression identical in form with (4.27) applies, it is possible to measure the relevant liquid–vapour surface tension. For the freezing transition no method of making a measurement of the ice–water interfacial free energy has yet been devised and theoretical estimates (McDonald, 1953) are very uncertain, though suggesting a value in the range 10–30 erg cm^{-2}. Since the theory leading to (4.31) involves a very questionable continuum approximation in any case, it is probably better to regard the effective σ as determined from nucleation experiments rather than to try to proceed in the opposite direction.

The experiments summarized in fig. 4.4 effectively measure the temperature at which the rate JV of nucleation of freezing in a drop of volume V is of order 1 s^{-1}. Using this criterion in the theory and taking $\sigma = 22$ erg cm^{-2} for best fit, we obtain the broken curve shown in fig. 4.4, in very good agreement with the lowest supercoolings found experimentally, which presumably refer to homogeneous nucleation. The value of σ is physically reasonable but should not be taken too seriously in view of the approximations involved. Because of the large value of the pre-exponential factor in (4.24), the nucleation rate is very critically dependent on the supercooling. Near -40 °C this rate decreases by a factor of twenty if the temperature is raised 1 degC so that homogeneous nucleation appears as a phenomenon with a quite sharply defined threshold which depends only slightly on the rate of cooling of the droplets in the experiment.

It is interesting to calculate the size of the critical embryo involved in nucleation. From (4.28) at -40 °C the critical radius is 11·3 Å, so that the embryo contains about 190 molecules. This result has interesting implications for the various flickering cluster theories of water structure. In order that the metastable supercooled state be maintained for reasonably long times at temperatures a few degrees above the nucleation threshold, it is necessary that formation of ice-like clusters of near-critical size should be a very rare event. Consideration of fig. 4.3 therefore requires that, whatever the structure of the clusters in the Némethy–Scheraga theory, they cannot be truly ice-like. This does not imply a rejection of the theory but simply a recognition that it is a model with only some of the features of the physical situation.

4.4. Heterogeneous nucleation

The theory outlined above explains in a most reasonable fashion the nucleation behaviour of extremely pure water. Most freezing, however, takes place with much less supercooling than this and it can only be concluded that some foreign influence aids the formation of ice-like clusters of more than critical size. Whilst it is known that strong electric fields or high-pressure sound waves have some effect on freezing, the mechanism is uncertain and we shall limit our discussion to the case of foreign materials which may be present in two forms—as suspended foreign particles and foreign surfaces or as dissolved molecules.

There is some evidence that dissolved molecules have some effects upon the cluster structure of liquid water and it is reasonable to expect this to have some influence on freezing behaviour. In particular, if some ion or molecule acts to stabilize a small ice-like cluster, this should aid nucleation, in competition with the general lowering of equilibrium freezing point produced by the solute concentration. Conversely, if the solute tends to disrupt the clusters or to encourage formation of clusters of a different form, then the probability of ice formation should be depressed. The basic work on the structural problem is that of Frank & Evans (1945). The more recent literature has been reviewed by Kavanau (1964) and by Samoilov (1965), and the effect on nucleation behaviour has been studied by Pruppacher (1963), who found that, for all materials tested, the structural effect of the solute was to increase slightly the supercooling required for nucleation. These increases ranged from 0 to 2 degC only and so are of little practical consequence.

The most important nucleating agents thus appear to be foreign surfaces and suspended particles of insoluble material. The way in which they act is fairly clear. If the molecular geometry and chemistry of an insoluble surface is suitable, then an ice-like cluster bounded on one side by this surface and on the others by liquid will have a lower free energy than an independent cluster and its growth will be facilitated. The general requirement for an efficient nucleating surface would therefore seem to be that the free energy of the interface which it forms with ice should be as low as possible. The discussion of this situation in terms of macroscopic parameters is much more nearly valid in this case than it was for homogeneous nucleation, for the critical embryo at about -10 °C contains thousands of molecules.

A simplified theory of heterogeneous nucleation can be worked out fairly easily for a foreign particle of arbitrary size and shape. If subscripts P, S and L refer to the particle, solid and liquid respectively, then the free energy involved in forming an ice embryo of volume V_S and surfaces A is

$$\Delta G = V_S \Delta G_v + \sigma_{SL} A_{SL} + (\sigma_{SP} - \sigma_{PL}) A_{SP}. \tag{4.32}$$

This expression can be simplified by defining an interface parameter

$$\mathfrak{m} = (\sigma_{PL} - \sigma_{SP})/\sigma_{SL} \tag{4.33}$$

with a possible range from -1 for a surface incompatible with ice to $+1$ for a surface which is 'wet' by ice. The free energy barrier to the growth of an ice cluster on the surface of a particle with a characteristic dimension R can then be written as

$$\Delta G_H^* = \Delta G^* f(\mathfrak{m}, R), \tag{4.34}$$

where ΔG^* is the free energy barrier to homogeneous nucleation, given by (4.31), and the function $f(\mathfrak{m}, R)$, which is always less than unity, depends in detail on the geometry of the particle and the habit of the embryo.

If the nucleating particle is a sphere of radius R and the ice embryo grows on it as a spherical cap, then (Fletcher, 1958)

$$f(\mathfrak{m}, R) = \frac{1}{2}\left\{1 + \left[\frac{1-\mathfrak{m}x}{h}\right]^3 + x^3\left[2 - 3\left(\frac{x-\mathfrak{m}}{h}\right) + \left(\frac{x-\mathfrak{m}}{h}\right)^3\right] + 3\mathfrak{m}x^2\left[\frac{x-\mathfrak{m}}{h} - 1\right]\right\}, \tag{4.35}$$

where $$h = (1 + x^2 - 2\mathfrak{m}x)^{\frac{1}{2}}, \quad x = R/r^* \tag{4.36}$$

and r^* is given by (4.28), (4.29) and (4.30). The function $f(\mathfrak{m}, R)$ has also been evaluated for a variety of more complex geometries, with nucleating particles ranging from disks to needles (Fletcher, 1963); the explicit forms are naturally different but the general behaviour is quite similar to that for the spherical case.

To derive an expression for nucleation rate per particle we use a procedure analogous to that for homogeneous nucleation discussed in the previous section. The main difference is that, instead of calculating a nucleation rate per unit volume, we calculate a rate per unit area of foreign substrate and then introduce the surface area of the particle, assumed to have homogeneous properties, to calculate the nucleation rate per particle. For a spherical particle of radius R cm the result is

$$J \sim 10^{20} R^2 \exp(-\Delta G_{\mathrm{H}}^*/kT) \quad \mathrm{s}^{-1}, \tag{4.37}$$

where ΔG_{H}^* is given by (4.34). The pre-exponential factor is smaller than that in (4.24) by a factor 10^7, the ratio of the number of water molecules per square centimetre of surface to the number per cubic centimetre of volume. With this result and the assumption of a value of 22 erg cm^{-2} for σ, as determined from homogeneous nucleation rates, we can calculate the temperature at which a spherical particle of radius R and surface parameter $\mathfrak{m}$ will nucleate freezing in 1 s. Curves showing the results of this calculation are given in fig. 4.8. It can be seen that nucleation efficiency depends strongly upon $\mathfrak{m}$ and that particles smaller than about 100 Å in radius are very much less active than are larger particles.

In practice, of course, one rarely deals with the properties of single particles but rather with distributions of many suspended particles of different sizes. In ordinary water there will also be particles of many different materials, but this added complication can be avoided in careful experiments. If the particle size distribution is known and a value for $\mathfrak{m}$ deduced—for example from the freezing threshold produced by large particles—then the statistical freezing behaviour of small drops can be determined. Comparisons carried out in this way give quite good agreement with experiment, though they do depend for this on the assumed value of $\mathfrak{m}$ and, in the case of experiments involving droplets suspended in air among fine smoke particles, an assessment of whether ice crystals are produced by the freezing of droplets or by direct sublimation onto the smoke particles from the vapour.

This theory, successful though it is, suffers from some major defects, the most important of which is the neglect of phenomena on a molecular level and the treatment of $\mathfrak{m}$ as a simple parameter characterizing all surfaces of the nucleating particle. Because of the

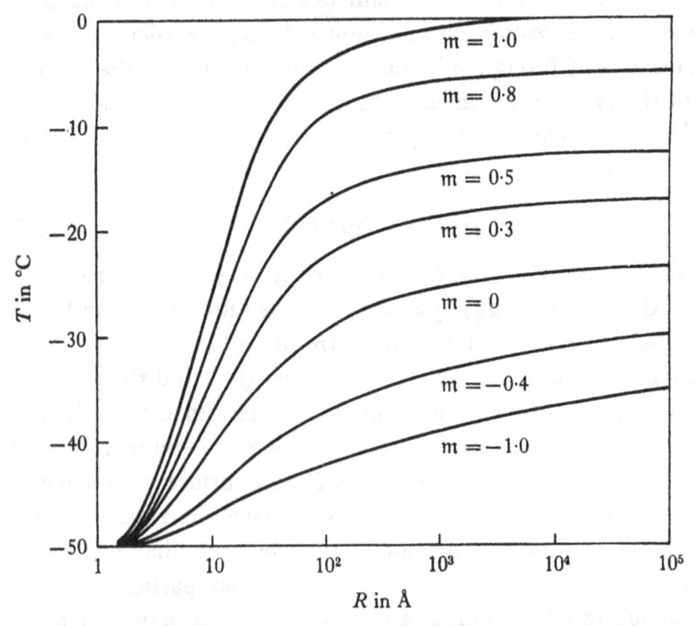

Fig. 4.8. Temperature T below which a particle of radius R and surface parameter $\mathfrak{m}$ will nucleate an ice crystal by freezing in less than 1 s (after Fletcher, 1958).

crystalline nature of ice, and generally of the particle as well, this must be an over-simplification and there is even the possibility that particularly active patches on the nucleating particles may play a decisive part in determining their behaviour (Edwards *et al.* 1962). Little has yet been achieved in the way of making an explicit identification of these active nucleation sites, though materials like silver iodide are known to show heterogeneous surface effects in water vapour adsorption which may be related to their ice nucleation behaviour (Zettlemoyer *et al.* 1963). Other recent experiments by Evans (1967*a*) on nucleation of ice under pressure or in salt solution suggest that the structure of the water monolayer in contact with the foreign substrate is of paramount importance and that there may be dominant nucleation and hysteresis effects occurring within this single layer. This is, indeed, reasonable on the basis of our general discussion of the nucleation problem, but no workable general theory has yet been developed which takes account of such effects at the molecular level. They are all, for the time being, lumped into the general surface parameter $\mathfrak{m}$.

A little progress has been made in evaluating the free energy of the ice/particle interface from first principles and some general rules for identifying efficient nuclei are discernible. Clearly the substrate should not introduce elastic strains into the ice embryo to any great extent for, though these can be accommodated in the theory (Turnbull & Vonnegut, 1952), they raise the free energy. This implies that the atomic fit between the substrate and some plane of an ice crystal should be as close as possible. Theoretical work (Read & Shockley, 1950; Van der Merwe, 1950, 1963; Fletcher, 1964, 1967; Fletcher & Adamson, 1966) shows that there are cusped energy minima when fairly general matching conditions apply across the interface and one would expect good nucleating agents to have crystal structures which come close to satisfying these conditions. Table 4.1 lists three materials which have good ice-nucleating properties and which have a close resemblance to ice in crystal parameters, though note the generalized character of the matching in the case of copper sulphide. When ice crystals grow from the vapour on these materials they do so in epitaxial relation, with one prominent crystal direction in the ice crystal parallel to another, not necessarily the same, in the substrate (Bryant *et al.* 1960).

This table lists only hexagonal crystals and only the best known of these. Silver iodide in the form of smoke particles a few hundred ångströms in average diameter, for example, is commonly used to seed clouds and induce ice formation at temperatures −10 to −15 °C instead of the −20 °C more characteristic of natural nuclei. Many materials which are not hexagonal and some which bear no obvious relation to ice structure, such as certain organic materials, are also quite efficient nuclei. Extensive experimental

TABLE 4.1. *Crystal relations for some ice-nucleating materials*

Material	a (Å)	c (Å)	$\frac{a-a_0}{a_0}$	$\frac{c-c_0}{c_0}$	Threshold (°C)
Ice	4·52	7·36	0	0	0
AgI	4·58	7·49	0·014	0·018	−4
PbI_2	4·54	6·86	0·005	−0·068	−6
CuS	3·80	16·43	0·028*	0·115†	−7

* [$\bar{2}$110] in ice along [$\bar{1}$010] in CuS surface.
† Two ice unit cells against one of CuS.

results have been given by Fukuta (1958, 1966), and by Mason & Van den Heuvel (1959).

Crystal structure is not, of course, the only factor influencing nucleation efficiency. For one thing, ice has a random orientation of molecular dipoles and a substrate which imposes any ordering on these will reduce the entropy of the ice molecules in the embryo and so increases its free energy (Fletcher, 1959). This effect appears to occur for silver iodide particles grown from solutions rich in either Ag^+ or I^- ions, these particles being much less active than those grown from isoelectric solutions (Edwards & Evans, 1962). The strength of the bonding between water molecules and the substrate will also certainly affect the nucleation efficiency. It is possible that the very high nucleation thresholds shown by some organic materials may be due to the OH groups exposed on fractured surfaces which presumably form energetically ideal anchorages for water molecules (Garten & Head, 1965).

The search for more and more efficient nuclei, both from the point of view of freezing threshold and of the number of active nuclei which can be produced from a given mass of material, is being pursued energetically because of the importance of ice crystals in cloud modification but further details of this search need not concern us here. A full discussion of the processes involved has been given by the author elsewhere (Fletcher, 1962*a*).

All this discussion has concerned the nucleation of ordinary Ice I_h from supercooled water but closely analogous nucleation processes are involved in the freezing of water under pressure to give one of the higher ices or in the conversion of one ice to another. The metastability of one phase in the region of stability of another is indicated by the dotted lines in the phase diagram in fig. 2.1. Foreign particles in suspension are usually very selective in their nucleation efficiency for the different ices, presumably because of geometrical relations across the interphase boundary, and this selectivity is sometimes great enough that one form of ice can be nucleated preferentially well into the stability region of another form (Evans, 1967*b*). There has, however, been little detailed study for the higher ices.

CHAPTER 5

CRYSTAL GROWTH

Ice crystals can grow in two simple distinct ways: either by the freezing of liquid water or by direct sublimation from the vapour phase. In each case the mechanisms which determine the rate and habit of growth are the transport of water molecules to the point of growth and their accommodation into the growing interface, together with the transport of latent heat away from this interface. Many different physical situations can occur, of course, but they are all controlled by these basic mechanisms.

If the system contains another component in addition to water, the situation can be much more complicated because crystal growth depends upon the transport of this component as well and there may also be competing processes occurring at the interface with the growing ice crystal. Some additional components, such as air in growth from the vapour, have a relatively small and simply understandable effect, but more complicated phenomena occur, for example, in the freezing of brine or sugar solutions. In this chapter we shall have very little to say about such cases but concentrate upon understanding the simpler systems.

5.1. Basic theory

The thermodynamic force driving any crystallization process is an excess of the chemical potential of molecules in the environment relative to its value in a bulk crystal. This excess may be specified as a supercooling, in the case of freezing, or as a supersaturation in the case of growth from the vapour or from solution. The total chemical potential excess, $\mu_\infty - \mu_S$, may be divided into two parts: a component $\mu_\infty - \mu_I$ which is the driving force for transport of water molecules through the environment to the growing interface, and a component $\mu_I - \mu_S$ which is necessary to achieve the incorporation of molecules into the crystal itself. This $\mu_I - \mu_S$ may be defined, quite generally but rather imprecisely, as the chemical potential excess at a distance of one mean free path length from the crystal surface. The transport problem, both for water molecules and for heat, differs considerably from one situation to another and must be worked out in detail for each case. The interface problem, however, can be divided conveniently into three simple possibilities which we now consider in turn. Those who require a more detailed discussion are referred to Burton *et al.* (1951), Cahn (1960) and Cahn *et al.* (1964).

In the first place, consider the growth of a perfect crystal face. If a single molecule is adsorbed, it is bound rather loosely to the crystal because of its small number of nearest neighbours and is likely to be freed again by thermal fluctuations. Two or more molecules form a more stable association and the formation of an island of this sort upon the crystal surface is a two-dimensional nucleation problem analogous to the nucleation problems treated in chapter 4. If a is the height of the island, r its radius and σ the surface free energy, assumed isotropic, then the total free energy for its formation is

$$\Delta G = \pi r^2 \Delta G_{\mathrm{V}}^{\mathrm{I}} + 2\pi r a \sigma, \tag{5.1}$$

where $\Delta G_{\mathrm{V}}^{\mathrm{I}} = -n_{\mathrm{S}}(\mu_{\mathrm{I}} - \mu_{\mathrm{S}})$. This expression has a maximum value

$$\Delta G^* = -\frac{\pi \sigma^2 a}{\Delta G_{\mathrm{V}}^{\mathrm{I}}} \tag{5.2}$$

for the critical radius

$$r^* = -\sigma / \Delta G_{\mathrm{V}}^{\mathrm{I}}. \tag{5.3}$$

Note the differences between these two relations for the present two-dimensional case and the analogous forms (4.27) and (4.28) for three-dimensional nucleation. The rate of advance of the interface is proportional to the rate of nucleation of new layers and therefore has the form

$$v = K \exp(-\Delta G^* / kT), \tag{5.4}$$

where K is a kinetic constant, determined in the case of freezing by considerations similar to those leading to (4.23) and in the case of sublimation by the same sort of method applied to vapour molecules. The general form of this growth law is shown by curve (*a*) of fig. 5.1.

For the specific case of freezing of supercooled water, (5.4), (5.2) and (4.29) lead to a growth law of the form

$$v = K \exp(-A' / \Delta T_{\mathrm{I}}), \tag{5.5}$$

where A' is a positive constant and ΔT_I is the supercooling at the interface. For growth from the vapour the chemical potential difference is

$$\mu_I - \mu_S = kT \ln (p_I/p_S), \tag{5.6}$$

where p_I is the partial pressure of water vapour at the interface and p_S the saturation vapour pressure over ice. Equation (5.4) thus leads to a growth law of the form

$$v = K' \exp[-A''/\ln(p_I/p_S)]. \tag{5.7}$$

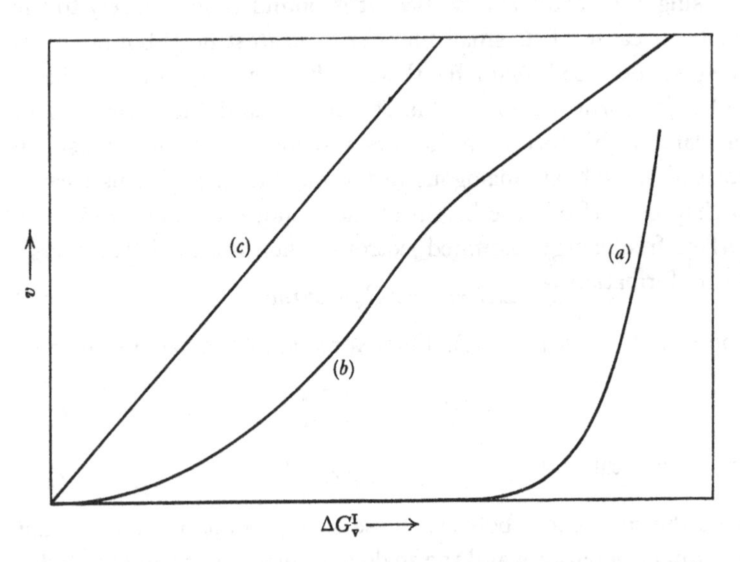

Fig. 5.1. Crystal growth velocity v as a function of driving free-energy difference ΔG_v^I for various mechanisms of interface advance. (*a*) Two-dimensional nucleation, (*b*) dislocation mechanism, (*c*) continuous advance of a rough interface.

The applicability of this sort of growth mechanism depends, however, upon the assumption that we are dealing with a perfect, smooth crystal surface. This is not often the case. Real crystal surfaces generally have imperfections of various kinds and, under certain circumstances, may even be rough on a molecular scale. The most important type of imperfection, from the point of view of crystal growth, is the screw dislocation. As was first pointed out by Frank (1949), a screw dislocation emerging from a crystal face provides a step on the surface to which atoms can be added continuously without causing it to disappear. The step is anchored to the point in the surface where the screw dislocation emerges and, as the step advances, it becomes twisted into a polygonal spiral, as investigated in detail by Burton *et al.* (1951). The rate of advance of the step at a place where its radius of curvature is r is simply proportional to the free energy excess of the neighbouring environment over that of molecules in the curved step

$$v_s(r) = -B(1-r^*/r)\Delta G_v^I, \tag{5.8}$$

where B is a constant and r^* is given by (5.3). In the steady state the spiral growth step is then, to a good approximation, of Archimedian form with co-ordinates (r, θ) given by

$$r = 2r^*\theta, \tag{5.9}$$

so that the distance between successive turns of the spiral is

$$d = 4\pi r^*. \tag{5.10}$$

The rate of forward growth of the crystal face on which the spiral forms is then approximately

$$v = \frac{v_s(\infty)}{d} = \frac{B}{4\pi\sigma}(\Delta G_v^I)^2. \tag{5.11}$$

Growth by this mechanism is thus proportional, in the case of freezing, to $(\Delta T_I)^2$ and, for sublimation from the vapour, to $[\ln(p_I/p_S)]^2$ and therefore approximately to the square of the supersaturation.

In vapour growth a new effect enters at high supersaturations. Transport of molecules to the growing steps is not primarily by direct collision; rather, the vapour molecules are adsorbed on the crystal surface and then diffuse across it to be added to the step. There is thus a characteristic diffusion distance upon the surface, determined by the mobility and residence time of the adsorbed molecules, from which each step can be considered to draw its molecules for growth. If the step separation d, determined by (5.10), were to fall below about twice this distance, then the growth rate of the two steps would be decreased. The step separation thus reaches a limiting value d' and the growth law becomes linear:

$$v = -(B/d')\Delta G_v^I. \tag{5.12}$$

The form of this growth law, quadratic tending to linear, is shown by curve (*b*) in fig. 5.1. The same law applies to surfaces from which emerge a whole distribution of screw dislocations, though the geometry is naturally more complex. The existence of these spiral growth terraces has, of course, been amply verified by observation.

For the third type of growth we recognize that interfaces, particularly those between a crystal and its melt, may not always be ideal plane or dislocated surfaces. Surfaces may, in fact, tend to be molecularly rough so that there is a certain diffuseness about their structure. This may represent an equilibrium situation (Burton *et al.* 1951; Jackson 1958) because the added entropy of the rough surface more than compensates for the extra energy caused by breaking nearest-neighbour bonds. In this case the criterion for roughness depends upon the ratio $\alpha L/kT$, where L is the latent heat and T the temperature of the phase transition, and α is the fraction of the total binding energy associated with bonding to other molecules in the surface layer. If this ratio is less than about 2·0 (which is the case for most metals in contact with their melts) rough interfaces result, while for higher values the interface is smooth. The effect which this transition has upon crystal habit is very marked (Jackson, 1966), rough interfaces leading to uniform and often curved crystal boundaries while smooth interfaces give rise to sharply defined crystal habit faces.

As well as surfaces which are rough in equilibrium, there is a general tendency for all surfaces to become rough when the free energy driving the crystal growth process is large. From (5.3) the radius of a critical surface embryo is $r^* = -\sigma/\Delta G_V^I$ and, to a rough approximation, if r^* is less than the radius of a single molecule $\frac{1}{2}a$ (which will happen if $-\Delta G_V^I > 2\sigma/a$), then such single adsorbed molecules will be stable and the surface will be rough. More refined calculations (Cahn, 1960) place the transition to a rough interface at

$$\Delta G_V^I = -\frac{\pi\sigma\eta}{a}, \tag{5.13}$$

where η is an additional parameter which measures the diffuseness of the equilibrium interface. η is essentially unity for an initially sharp interface and decreases rapidly as the diffuseness increases, its approximate value for an interface spread over z molecular layers being

$$\eta \simeq \tfrac{1}{16}\pi^4 z^3 \exp\left(-\tfrac{1}{2}\pi^2 z\right). \tag{5.14}$$

If a crystal has a rough surface then there is no nucleation barrier to be overcome in crystal growth and the rate can be very simply calculated. If the chemical potential at one mean free path length l from the interface is μ_I, then the thermodynamic force driving crystal growth is $(\mu_I - \mu_S)/l$ and the generalized mobility for molecules joining the crystal is D/kT, where D is the coefficient of self-diffusion for molecules in the parent phase. From this the growth rate is

$$v = \frac{D}{kT}\frac{\mu_I - \mu_S}{l}\frac{n'}{n_S} = \frac{Dn'}{kTn_S^2 l}\Delta G_V^I, \tag{5.15}$$

where n' is the concentration of molecules in the parent phase and n_S that in the crystal. When, as is usually the case, the transition is one of freezing, then $n' \simeq n_S$ and l is of the order of, but rather less than, the atomic spacing a, so that (5.15) becomes, using (4.29),

$$v \simeq \frac{D\Delta S_V \Delta T_1}{kTn_S a}. \tag{5.16}$$

This simple linear behaviour is shown as (*c*) in fig. 5.1 and also appears as the limit of all other curves in the figure for very high driving energies.

Determination of the kinetics of growth at the interface is, of course, only one part of the description of the kinetics of crystal growth. The component $\mu_\infty - \mu_I$ of the driving potential must, if there is any diffusive flow, drive the migration of molecules towards the growing interface while, at the same time, all the potentials μ_S, μ_I and μ_∞ are affected by the necessity for transport of latent heat by means of temperature gradients. Both these fluxes depend linearly upon the driving forces concerned. In different experimental situations any one of these mechanisms may be dominant in determining the kinetics of crystal growth and the problem must be carefully analysed before conclusions about the interfacial growth mechanism can be reached from experimental data.

The only case which can be treated at all simply is that in which the crystal growth is so slow that the diffusion equation governing matter transport and the formally identical equation for heat flow

$$\frac{\partial \phi}{\partial t} = \beta \nabla^2 \phi, \tag{5.17}$$

where ϕ is either molecular concentration or temperature and β is either diffusion coefficient or thermal diffusivity, reduces approximately to Laplace's equation

$$\nabla^2 \phi = 0. \tag{5.18}$$

As an example, consider the growth of an isolated isothermal crystal in which all crystal faces grow similarly. The boundary conditions are then $\phi = \phi_{\mathrm{I}}$ at the interface and $\phi = \phi_\infty$ at infinity. The electrostatic analogy is clear and the molecular and heat fluxes to the growing crystal can be written

$$J_n = 4\pi CD(n_\infty - n_{\mathrm{I}}) \tag{5.19}$$

and

$$J_T = 4\pi C\kappa(T_\infty - T_{\mathrm{I}}), \tag{5.20}$$

where D is the molecular diffusion coefficient and n the number density of molecules in the environment and κ its thermal conductivity. C is a factor dependent on the size and shape of the growing crystal and is numerically equal to its electrostatic capacity expressed in electrostatic units (thus, for a sphere of radius r, $C = r$). The two fluxes given above are linked by the relation

$$J_T = -(L/N)J_n, \tag{5.21}$$

where L is the latent heat of crystallization per mole and N is $6{\cdot}02 \times 10^{23}$.

For the freezing of a pure supercooled liquid (5·19) does not apply and (5·20) and (5.21) lead to a growth rate which depends linearly upon $(T_\infty - T_{\mathrm{I}})$. The rate of growth at the interface, however, depends on $(T_{\mathrm{I}} - T_{\mathrm{S}})$ according to (5.5), (5.11) or (5.16), and the necessity for equality of these two rates fixes T_{I} and hence the over-all growth kinetics. Similar considerations apply to growth from solution or from the vapour but in these cases (5.19) must be taken into account as well. For more complicated situations, such as the growth of crystals in tubes, the same general approach can be used though the boundary conditions upon the solution of (5.18) are necessarily different. If the crystal growth is so rapid that the full equations (5.17) must be used, then there is not generally any simple form of solution.

5.2. Freezing of water

Before considering the kinetics of freezing, it is worth while to say a little about the morphology of ice crystals grown in this way. This description is, unfortunately, not simple and the shapes of the crystals produced often depend more upon the heat flow conditions imposed by the experiment than upon any intrinsic properties of ice.

In situations in which the liquid water is slightly above freezing temperature and heat is withdrawn through the ice—as, for example, in the freezing of a pond cooled by radiation to the night sky—the interface between the ice and water is very nearly smooth, with small grooves marking the grain boundaries (Harrison & Tiller, 1963). The preferred grain orientation is usually either with the *c*-axis normal to the interface (because thermal conductivity is about 5 per cent greater parallel to the *c*-axis than perpendicular to it) or with the *c*-axis parallel to the interface (because, as we shall see, the growth velocity is greater perpendicular to the *c*-axis). The texture developed depends on many things (Knight, 1966), among them the nucleation process in the thin layer of supercooled water initially formed at the surface. Crystals with *c*-axis normal to the surface develop as thin, almost circular disks, while those with *c*-axis in the surface grow as long needles parallel to an *a*-axis fast growth direction. Details of subsequent development need not concern us here.

When ice is growing in supercooled water the direction of major heat flow is generally from the crystal to the liquid. This may lead to an unstable situation, since any part of the crystal interface which gets ahead of neighbouring regions is in a more favourable position to dissipate latent heat of crystallization and so its growth rate is enhanced. This situation thus often leads to dendritic (or tree-like) crystal growth with dendrite arms branching out into the liquid. At temperatures above about $-2{\cdot}5$ °C we find plane stellar dendrites or dendrite sheets and, when a small ice crystal is introduced into a volume of supercooled water at temperature below −2·5 °C, the growth pattern of the dendrites is such as to give twelve primary growth directions, resulting in a double

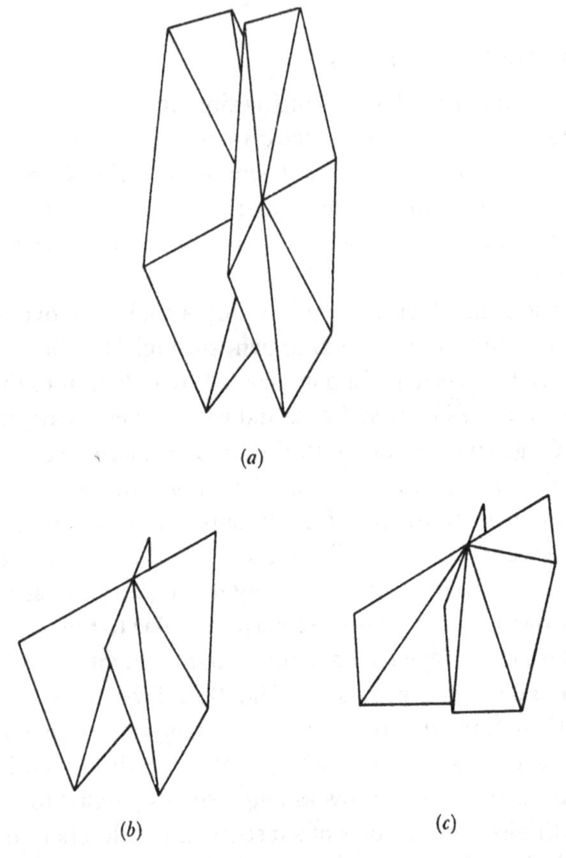

Fig. 5.2. Schematic representation of ice structures growing in supercooled water at temperatures below −3 °C. (*a*) Complete structure with twelve primary growth directions emanating from the central point of nucleation and symmetrically placed about the basal plane of the seed crystal; (*b*) and (*c*) typical structures for nucleation at the liquid surface (Macklin & Ryan, 1965).

hollow pyramid as shown in fig. 5.2 and plate 2. These structures have been extensively studied by Macklin & Ryan (1965, 1966) and by Pruppacher (1967*a*), who find that the dendrite sheets are not rational crystallographic planes but are separated by an angle

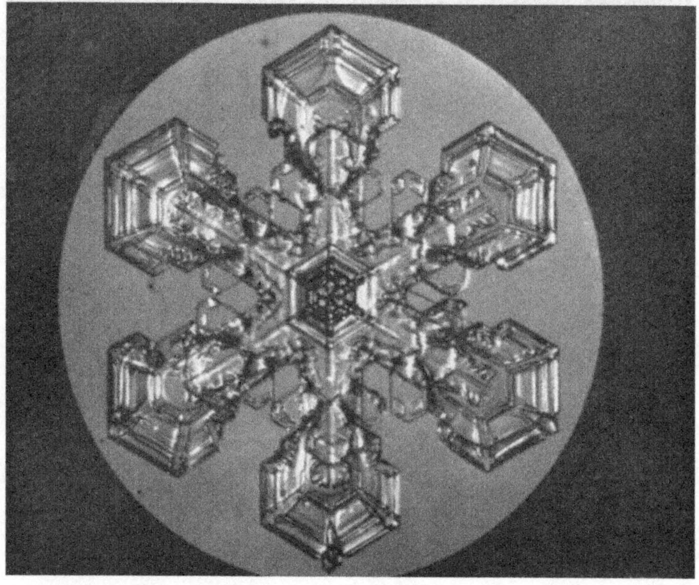

1 Photomicrographs of typical natural snow crystals, showing almost perfect hexagonal symmetry in their habit. The crystals are, in each case, approximately 2 mm in diameter (Magono & Lee, 1966).

(*Facing p. 112*)

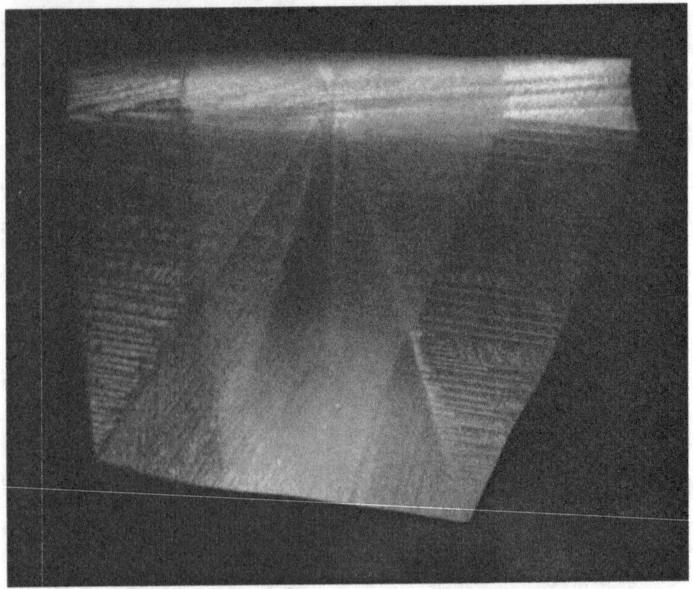

2 Two views of a typical growth form of ice in supercooled water. The upper photograph is taken looking down on the water surface and the lower one parallel to the surface, which cuts across the top of the picture. The crystal photographed is approximately 2 cm across (Macklin & Ryan, 1965).

which depends upon the supercooling of the bulk liquid as shown in fig. 5.3. The growth mechanism is not entirely clear but may involve some sort of stepped or segmented growth whose pattern depends upon the different temperature variation of growth velocities parallel to *c*- and *a*- axes. When the bath supercooling exceeds about 6 degC the ice begins to develop secondary growth planes roughly parallel to the primary axes and a more compact spongy structure is formed.

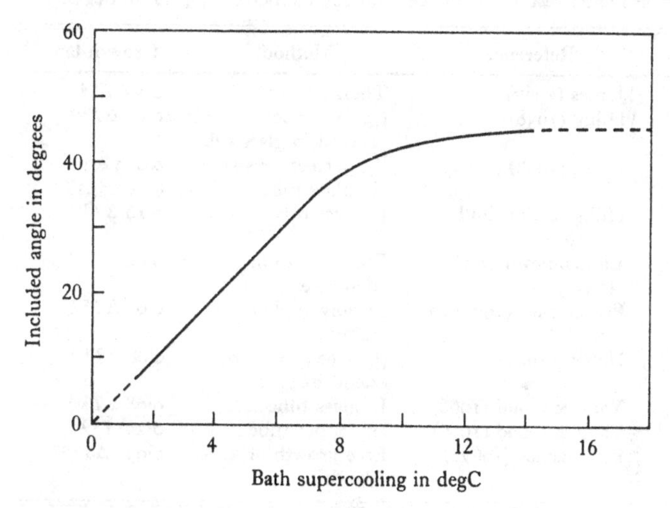

Fig. 5.3. Variation with supercooling of the angle between primary growth directions of the ice structures of fig. 5.2 as determined by Macklin & Ryan (1965, 1966) and Pruppacher (1967*a*). The angular spread of experimental results is about $\pm 4°$.

If the water used for the experiments contains appreciable amounts of dissolved impurity, then the growth structures may be considerably modified. Solute rejected from the advancing crystal increases the concentration near the growing interface and may lead to the familiar metallurgical situation of constitutional supercooling and ridged, cellular or dendritic interfaces (Harrison & Tiller, 1963). The structures developed in supercooled solutions are generally similar to those in pure water, though the angular dependence is slightly modified (Macklin & Ryan, 1966; Lindenmeyer & Chalmers, 1966*a*; Pruppacher, 1967*a*).

A final case worthy of note is the growth of ice upon foreign surfaces, generally when the water is present as a thin supercooled layer. The major effect of the foreign surface, if its thermal conductivity is greater than that of water, is to aid transfer of heat away from the interface. The dendritic growths which result may be curved and the growth mechanism is, as yet, far from clear (Knight, 1962; Lindenmeyer & Chalmers, 1966*a*).

TABLE 5.1. *Growth velocity of ice crystals*

Experimental results in cm sec^{-1} for total supercooling ΔT in degC.

Curve	Reference	Method	Growth law
1	James (1967)	Thermal wave	$0.1\ \Delta T^{1.3}$
2	Hillig (1958)	$\Vert_c$, imperfect crystal in glass tube	$0.01\ \Delta T^{1.2}$
3	Hillig (1958)	$\Vert_c$, perfect crystal in glass tube	$0.003\ \exp(-0.25/\Delta T)$
4	Hillig & Turnbull (1956)	In glass tube	$0.16\ \Delta T^{1.7}$
5	Lindenmeyer *et al.* (1957)	Free growth of dendrite	$0.02\ \Delta T^{2.2}$
6	Pruppacher (1967*b*)	In polyethylene tube	$0.05\ \Delta T^{2.1}$
7	Hallett (1964)	$\Vert_a$, free growth of dendrite	$0.08\ \Delta T^{1.9}$
8	Yang & Good (1966)	In glass tube	$0.08\ \Delta T^{1.8}$
9	Yang & Good (1966)	In copper tube	$0.2\ \Delta T^{2.1}$
10	Pruppacher (1967*c*)	Free growth of dendrite	$0.035\ \Delta T^{2.22}$

Turning now to the kinetics of growth, we recall that the roughness of the interface is controlled by the parameter $\alpha L/kT$. For the most densely packed planes of ice, the fraction α of bonds in the surface layer is $\frac{3}{4}$ so that $\alpha L/kT$ is very close to the critical value 2·0 and nothing definite can be said about the roughness of the ice/water interface. The occurrence of dendritic forms and both smooth and faceted interfaces under different conditions is consistent with the transitional value of this parameter (Jackson, 1966).

Measurements of crystallization velocity fall into two general classes: those in which free dendrite growth is measured and those which confine the growing ice in a tube which is placed in a bath of some other liquid. The first has the prime advantage of eliminating any substrate effects but the estimation of the interface temperature is rather difficult. The various results found by different workers, annotated as to method, are displayed in fig. 5.4 and table 5.1, where growth velocity is given as a function simply of total supercooling ΔT. With the exception of curves 2 and 3,

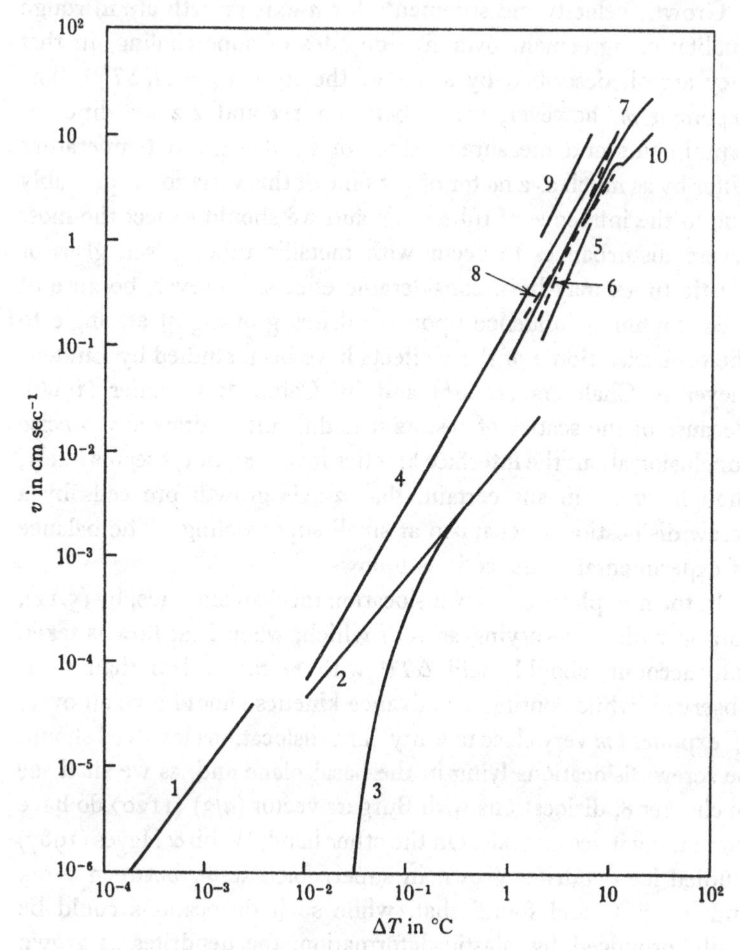

Fig. 5.4. Experimental results for the growth velocity v of ice crystals in terms of total supercooling ΔT. References are to table 5.1.

which are for growth parallel to the c-axis, all curves are either specifically for growth in an a-axis direction or, when no orientation was specified, probably apply to this fastest growth direction. Curve 1, for extremely low growth rates, was determined by a thermal wave method in which the change in amplitude of a thermal disturbance crossing the solid/liquid interface is measured and this change is attributed to absorption of latent heat caused by interface motion (Kramer & Tiller, 1962).

Growth velocity measurements for *a*-axis growth are in rough qualitative agreement over five decades of supercooling, in that they are all described by a law of the form $v_a = A(\Delta T)^m$. The exponent m, however, varies between 1·3 and 2·2 for different experiments and measured values of v_a at a given temperature differ by as much as a factor of 5. Some of this variation is probably due to the influence of tube walls and we should expect the most severe disturbances to occur with metallic tubes. Even glass or plastic tubes may have considerable effects, however, because of their confining influence upon dendrites growing at an angle to the tube axis. Some of these effects have been studied by Lindenmeyer & Chalmers (1966*b*) and by Camp & Creamer (1966). Because of the scatter of results it is difficult to draw any precise conclusion about the interface kinetics involved, but it seems likely, though by no means certain, that *a*-axis growth proceeds by a screw dislocation mechanism at small supercoolings. The balance of experimental evidence is as follows.

In the first place, a screw dislocation mechanism leads, by (5.11), to a growth rate varying as ΔT_I^2 which, when heat flow is taken into account, should yield ΔT^m with m rather less than 2, as observed, while continuous advance kinetics should give an overall exponent m very close to unity. The dislocations involved should be screw dislocations lying in the basal plane and, as we shall see in chapter 8, dislocations with Burgers vector $(a/3)\langle 11\bar{2}0\rangle$ do have low energy in ice crystals. On the other hand, Webb & Hayes (1967) studied ice dendrites grown in supercooled water between −0·5 and −1·5 °C and found that, while such dislocations could be easily produced by plastic deformation, the dendrites as grown were almost completely dislocation-free.

Secondly, if the ice/water interface is smooth at low supercoolings so that growth requires a dislocation mechanism, then from our discussions we should expect a transition to continuous growth at some larger supercooling. Hallett's results (curve 7) extend to −20 °C without any very significant change in growth law, though perhaps a slight decrease in the exponent. Pruppacher (1967*c*) (curve 10), on the other hand, finds a transition to a linear law for $\Delta T > 9$ degC for free dendrite growth, which appears to indicate the onset of a continuous growth mechanism. Taking heat transfer effects into account, the interface supercooling at which the transition takes place is $\Delta T_I \simeq 2$ degC. This can be made to agree with the prediction of (5.13) if the parameter η is about 10^{-2}, which by (5.14) corresponds to a diffuseness of about two molecular layers at the interface. This seems physically reasonable but the variation of η with interface diffuseness is so sharp that the experimental evidence is in no way crucial.

Turning now to growth parallel to the *c*-axis, the only experimental results are those of Hillig (1958) shown as curves 2 and 3 in fig. 5.4. Crystals of fairly high perfection apparently grow by a mechanism of two-dimensional nucleation and curve 3, when corrected for the difference between bath supercooling and interface supercooling, gives $v_c = 0{\cdot}03 \exp(-0{\cdot}35/\Delta T_I)$ cm s^{-1}. This is of the form of (5.5) but the numerical value of the coefficient in the exponent implies the rather small value of 6 erg cm^{-2} for the interfacial free energy, compared with the value near 22 erg cm^{-2} derived in chapter 4 from studies of homogeneous nucleation. It may be that the (0001) surface is also slightly diffuse, resulting in a lowered edge energy for steps, or there may be a large variation of the interfacial free energy with temperature.

Since perfect basal planes are unable to grow by a continuous mechanism, the basal face of a damaged crystal is presumably smooth except for steps produced by dislocations. The $\Delta T^{1{\cdot}2}$ growth behaviour shown by such imperfect crystals is therefore likely to be some modification of the ΔT_I^2 law for dislocation growth rather than a modification of the ΔT_I behaviour for continuous growth. Certainly screw dislocations have been found on basal faces (Muguruma, 1961), though this does not necessarily imply their part in the growth process. Hillig's observations of interface structure appear to indicate that the glass tube has little effect upon crystal growth in his case but it cannot be entirely eliminated as a disturbing influence on the kinetics.

It is interesting to note that, from curves 2 and 4 of fig. 5.4, taken for *c*-axis and *a*-axis growth under similar conditions, the ratio v_c/v_a decreases as ΔT increases. This is opposite to the tendency suggested by the simplest interpretation of the growth structures of fig. 5.2 as due to variation of this ratio, since fig. 5.3 shows that the angle of separation, essentially $2v_c/v_a$, increases with ΔT. The growth illustrated in fig. 5.2 takes place, however, at much greater supercoolings than those covered by the measurements of Hillig, so it is possible that some change in *c*-axis growth mechanism occurs for $\Delta T > 2$ degC.

The growth of ice crystals from aqueous solution is a subject whose ramifications would take us too far afield to be discussed here. Ionic solutes in concentrations up to 0.1 M do not greatly affect either the growth forms or crystallization kinetics (Lindenmeyer & Chalmers, 1966*b*; Pruppacher 1967*a*, *b*), though the actual crystallization velocity at a given supercooling (relative to the equilibrium freezing point of the solution) may be changed by a factor of 2 or so. Higher solute concentrations tend to depress the crystallization velocity more severely. In sugar solutions, however, when the concentration of sugar exceeds about 10^{-2} M, the interface kinetics appear to become linear in ΔT (Lindenmeyer & Chalmers, 1966*b*), indicating a change in interface structure.

To conclude this section we note that single crystals to be used in studying the physical properties of ice are most conveniently prepared by growth from the melt. Impurities or doping agents can be easily incorporated by adding them to the water, taking account of the segregation which occurs on freezing. Any of the usual crystal growth methods can, with simple modifications, be applied to ice. The general principles are discussed by Brice (1965) and specific applications to ice have been given by Jona & Scherrer (1952) and by Jaccard (1959).

The distribution of impurities between an aqueous solution and a growing ice crystal follows the same laws as for other crystal materials. The detailed behaviour depends upon the nature and concentration of other materials present but, for freezing of dilute solutions, the parameter of chief concern is the segregation coefficient, which gives the ratio of solute concentration in the crystal to that in the liquid from which it is growing. For very slow growth rates this coefficient can be determined from the phase diagram but in general it depends upon growth rate and stirring conditions in the solution. Most of the distribution coefficients which have been determined for ice relate to impurities which enter the ice structure substitutionally, replacing a water molecule. All these materials are preferentially rejected by the growing crystal, the segregation coefficient for hydrogen fluoride or ammonia being of order 10^{-2} but increasing with increasing growth rate. The salt ammonium fluoride has a rather larger solubility in ice and a segregation coefficient between 10^{-2} and 10^{-1} because the disturbing effects of its two components are in opposite senses and tend to cancel. Details of this work are given by Decroly & Jaccard (1957), by Jaccard & Levi (1961) and by Jaccard (1966). Some impurities may, alternatively, enter the crystal interstitially or, for very rapid freezing and dendritic growth, pockets of concentrated solution may become trapped within the growing crystal. These solution pockets are very important for many of the properties of sea ice (Pounder, 1965, chapter 2) but will not concern us here.

5.3. Growth from the vapour

The growth of ice crystals from the vapour is a familiar phenomenon since a major part of the growth of snow flakes occurs in this way. The beautifully symmetric growth forms of snow crystals, like those shown in plate 1, are well known and it is often supposed that all natural snowflakes are of this form. This is actually far from being the case, particularly with crystals exceeding 1 mm or so in diameter, which may often be broken, malformed or rimed with frozen water droplets. Most of these structures, though meteorologically important, can for our present purposes be dismissed as growth accidents and we may concentrate upon the symmetric forms as representing ideal undisturbed growth from the vapour. Even within these symmetric forms a great variety of habits is possible and these have been classified, together with the irregular forms, by Nakaya (1954) and the classification brought up to date by Magono & Lee (1966). These habits range from long thin needles through columns to plates and each form may be simple or dendritic. As we shall see later, the simpler habits can all be related to the environment in which the crystal grew. In addition there are many mixed forms, such as columns tipped with plates, representing crystals whose environment altered during the course of their growth. All these habits give useful meteorological information about conditions inside clouds but we shall concentrate rather upon the physics of the growth processes themselves.

The parameters affecting the growth habits of ice crystals are fairly clearly the temperature and the supersaturation of the vapour. Contaminants may also have a very marked effect but this is not intrinsic to the behaviour of ice. Nakaya (1954) was among the first to investigate the effects of temperature and supersaturation upon ice crystal habit, using for this purpose a chamber in which ice crystals grew upon a fine fibre and in which the temperature and supersaturation could be varied with a moderate amount of independence. These experiments were greatly refined by Hallett & Mason (1958) using a diffusion chamber to define the crystal environment. In such a chamber the side-walls are of insulating material and the top plate, which is kept moist, is at a higher temperature than the bottom plate. There is thus a vertical gradient of both temperature and vapour density and the non-linear relation between the equilibrium values of these two quantities leads to a supersaturation near the middle of the chamber.

The results of these experiments are shown in fig. 5.5, known as a Nakaya diagram, in which are plotted the regions of occurrence of different crystal habits. A striking feature of this diagram is the vertical lines dividing temperature regions in which the growth is alternately axial, leading to long needles orientated parallel to the *c*-axis, and lateral, yielding flat hexagonal plates. The transitions between the different growth regimes are very sharp, extending over less than 1 degC, the axial ratio c/a of the crystals changing discontinuously from perhaps 10 for columns and even 100 for needles to values from 0·1 to 0·01 for plates. It is also worth noting that transitional forms, such as plates growing on the ends of needles, can be produced if the crystals are transferred from one temperature regime to another during growth. The temperature dependence of crystal habit is thus extraordinarily specific.

The variation of habit with supersaturation is rather less spectacular and more easily understood. At high supersaturations, and thus at high growth rates, the crystal forms become more complex and tend to dendritic development. The dendrites may be planar embellishments of the crystal form, as shown in plate 1, or may lead to scrolls, cups or hollow prisms in appropriate temperature ranges.

Since it is the vapour density gradient which determines crystal growth rates, it is perhaps more meaningful to display the experi-

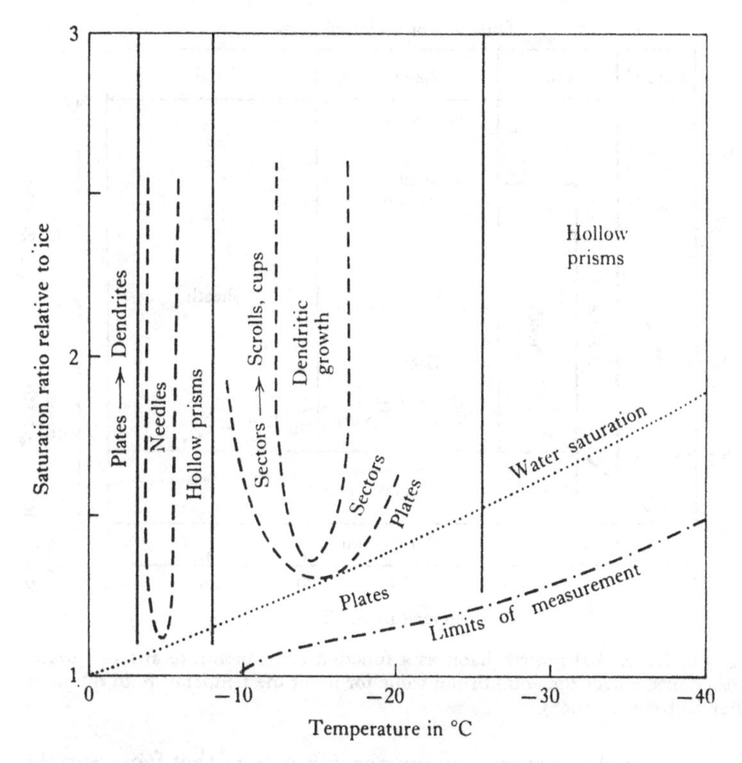

Fig. 5.5. The growth habits of ice crystals, grown in a diffusion chamber, in relation to temperature and supersaturation (Hallett & Mason, 1958).

mental results in terms of excess vapour density rather than supersaturation. This is done in fig. 5.6, due to Kobayashi (1958) who also extended the measurements to very low growth rates and found there quasi-equilibrium growth habits in the form of short columns. As the vapour density excess is increased, growth tends to be concentrated at crystal edges and then at corners, as is to be expected if transport on the crystal surface is limited, since molecules tend to be deposited where the vapour density gradient is largest.

If the growth of an ice crystal from the vapour were a simple near-equilibrium process, then the resulting crystal habit could be determined by Wulff's theorem (Wulff, 1901) which states that, in equilibrium, the distance of any face from the centre of the crystal

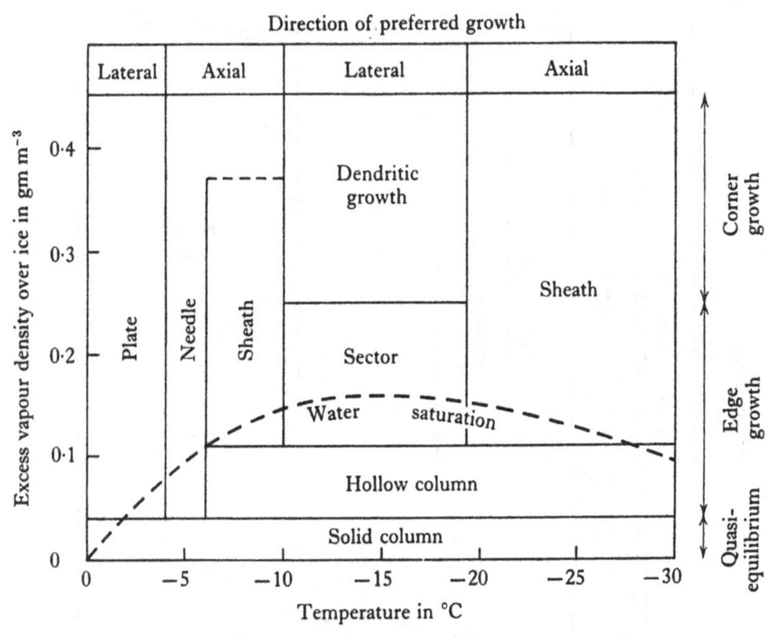

Fig. 5.6. Ice crystal growth habit as a function of temperature and of vapour density excess over the equilibrium value for ice at the temperature in question (after Kobayashi, 1958).

is proportional to the free energy per unit area of that face. Simple bond counting suggests that the energy of a prism face is about 6 per cent greater than that of a basal face so that, neglecting entropy differences, Wulff's theorem predicts that the equilibrium form should be a hexagonal prism, the ratio of whose axial length to maximum transverse dimension is 0·81. This is approximately the habit found by Kobayashi (1958) for very slow growth. It is obvious that the extreme forms found under more normal growth conditions cannot be explained by simple variations in surface free energy but must rather be due to kinetic effects associated with the rate of transfer of water molecules across the surface of the growing crystal and their inclusion into its lattice.

Details of the surface phenomena responsible for these habit changes are not yet understood but important basic information upon the mechanism of ice crystal growth from the vapour has been obtained from the study of crystals growing upon foreign substrates. Such crystals begin their life after a nucleation event of the type discussed in connexion with freezing in chapter 4. The free energy barrier to nucleation from the vapour, as given by (4.34) and (4.27), is larger than that for nucleation from the liquid phase because

$$\sigma_{\text{ice/vapour}} \simeq 100 \text{ erg cm}^{-2}$$

compared with

$$\sigma_{\text{ice/water}} \simeq 22 \text{ erg cm}^{-2}.$$

Thus Bryant *et al.* (1960) find that a supersaturation of 12 per cent with respect to ice is required to nucleate ice crystals directly from the vapour on a silver iodide substrate. Since for a vapour pressure p

$$\Delta G_{\text{v}} = -n_{\text{S}} kT \ln(p/p_{\text{S}}), \tag{5.22}$$

where p_{S} is the saturation vapour pressure over ice, this corresponds to $\Delta G_{\text{v}} = -0{\cdot}12 n_{\text{S}} kT$. On the other hand, silver iodide nucleates freezing at -4 °C which, by (4.30), corresponds to $\Delta G_{\text{v}} = -0{\cdot}04 n_{\text{S}} kT$. It is thus thermodynamically easier, above -12 °C, to first form water droplets by condensation upon the substrate (which requires a supersaturation of less than 3 per cent relative to liquid water) and convert these to ice by a freezing process. Below -12 °C, since water saturation is achieved only at $p/p_{\text{S}} \simeq 1{\cdot}01 \Delta T$, ice can nucleate directly from the vapour.

The conditions for a good nucleating substrate are similar to those discussed for freezing, and ice crystal growth generally occurs epitaxially. A case of particular interest is that of the growth of ice upon the mineral covellite (CuS), for which the ice grows as uniform plates in close contact with the substrate and so thin that interference colours can easily be observed. This allows a close quantitative study of crystal growth to be made and, in particular, the propagation across the surface of growth steps of known height can be followed in detail. These phenomena are illustrated in the paper by Bryant *et al.* (1960) and the measurements are analysed in subsequent papers by Hallett (1961) and by Mason *et al.* (1963).

The observations show that growth in the c-axis direction is not continuous but occurs by step propagation across the basal plane up to vapour density excesses of at least 3×10^{-7} g cm^{-3} within the experimental chamber, such density excesses being reached at -20 °C when the vapour saturation ratio p/p_S is about 1.5. This is consistent with the discussion of chapter 4 which predicts, since $\alpha L/kT \simeq 15$ for the crystal-to-vapour transition, that the equilibrium interface should be smooth. From (5.2) and (5.4) we also find that the rate of nucleation of new layers of unit step height should be small for p/p_S less than about 1.5 at the interface, which is again consistent with observations.

The motion of the growth steps is consistent with the hypothesis that vapour molecules are adsorbed in equilibrium concentration upon the flat basal surface and that those within some characteristic surface diffusion distance l_s can migrate to the step and become incorporated in the crystal. If the step height is $h \ll l_s$ then the rate of step motion should be

$$v = B'l_s/h, \tag{5.23}$$

where B' depends upon p/p_S and T but is independent of h. This relation (5.23) was verified experimentally, h being determined from interference colours when in the range 200–1000 Å, corresponding to some hundreds of molecular layers. Further, by examining the separation at which interaction between steps occurs, the absolute value of l_s at different temperatures was determined. The results are not highly accurate and there is some ambiguity in interpretation because of the part played by supersaturation, but the temperature dependence of l_s, as shown in fig. 5.7, is remarkable (Mason *et al.* 1963). The variation of step velocity with temperature is of the same form (Hallett, 1961).

The theory of surface diffusion and step motion for simple crystals is well known (Burton *et al.* 1951). From the formula due to Einstein, the surface diffusion length l_s is given in terms of surface diffusion coefficient D_s and surface adsorption lifetime τ_s by

$$l_s = (D_s\tau_s)^{\frac{1}{2}}. \tag{5.24}$$

Now if H_s^t is the activation energy for transport by surface diffusion, which is approximately equal to one hydrogen-bond breaking energy, and a is the jump distance for surface molecules, then

$$D_s = a^2 \nu \exp(-H_s^t/kT), \tag{5.25}$$

where ν is of the order of an atomic vibration frequency. Similarly the surface adsorption lifetime is given by

$$1/\tau_s = \nu \exp(-H_s^a/kT), \tag{5.26}$$

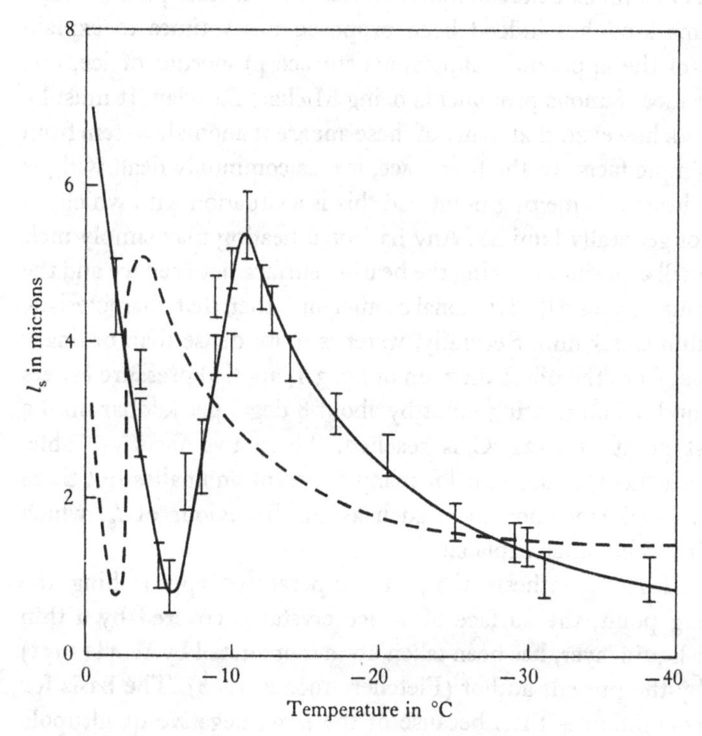

Fig. 5.7. The variation with temperature of the surface diffusion length l_s for water molecules on the basal face of an ice crystal. The broken curve shows the conjectured behaviour of l_s on a prism face (Mason *et al.* 1963).

where H_s^a is the energy binding an adsorbed molecule to the surface, which is probably about twice the hydrogen-bond total energy. Thus

$$l_s \simeq a \exp[(H_s^a - H_s^t)/2kT], \tag{5.27}$$

which indicates that l_s is of the order of a few microns and decreases as the temperature is raised. The magnitude is of the observed order but the temperature dependence is not that found for ice.

Some of this discrepancy may be resolved by the realization that, under the conditions of the experiment, the amount of adsorption may not be small (Hobbs & Scott, 1965) as assumed in the simple theory, but it seems likely that the possibility of some sort of structural change in the surface must be invoked to explain the very complex behaviour above −12 °C. A surface phase change of some kind has indeed been proposed many times to explain some of the apparently anomalous surface properties of ice, one of the most famous proponents being Michael Faraday. It must be realized, however, that many of these apparent anomalies stem from two simple facts. In the first place, ice, as commonly dealt with, is very close to its melting point and this is a situation with which we are not generally familiar. Any frictional heating may simply melt the small asperities forming the bearing surface between ice and the other material and the frictional coefficient is then that characteristic of a thin water film. Secondly, water is more dense than ordinary ice and, from the phase diagram of fig. 3.1, applied pressure lowers the equilibrium melting point by about 8 degC per kilobar until a lowest point of −22 °C is reached. These two facts are able, between them, to account for many apparent anomalies but there are residual real anomalies, such as the behaviour of l_s, which require some other explanation.

Faraday's hypothesis that, at temperatures approaching the melting point, the surface of an ice crystal is covered by a thin quasi-liquid layer, has been taken up in more detail by Weyl (1951) and by the present author (Fletcher, 1962*b*, 1968). The basis for this contention is that, because of the large negative quadrupole moments of the water molecule, it is energetically favourable in liquid water to have a substantial degree of molecular orientation near the surface, decaying to randomness in the interior with a relaxation length of about 10 Å, as characteristic of the bonding structure of water. The free energy penalty paid for dipole–dipole interactions and for loss of entropy is more than balanced by the energy gained from dipole–quadrupole interactions, and the low energy state is that with the protons of the surface molecules directed preferentially out of the surface. (The less realistic model used by

Fletcher in 1962 suggested the opposite orientation.) At a crystalline ice surface such molecular orientation cannot occur because of the complete bonding within the crystal. However, it may be thermodynamically favourable to have a thin disordered quasi-liquid layer on the surface at temperatures close to the melting point so that surface orientation can occur and decay within the surface layer to the random value characteristic of bulk ice. The published calculations based on this model (Fletcher, 1968) are necessarily speculative but they do predict the existence of a quasi-liquid surface above about -6 °C with thickness of a few molecular layers, increasing to about ten layers at -1 °C. Though these figures cannot be relied upon, they do suggest that some sort of surface phase change may occur within about 10 degC of the melting point and this may easily be responsible for the complicated behaviour of l_s, together with some others of the apparently anomalous surface properties of ice (Jellinek, 1961).

Returning to the problem of ice crystal habit, this sharp dependence of surface diffusion properties upon temperature provides a possible mechanism, especially since we have agreed that such extreme habit changes must be the result of surface kinetics rather than of equilibrium energetics. So far it has not proved possible to perform an analogous growth experiment on the prism faces of ice crystals but Mason *et al.* (1963) have conjectured that, if such faces show a similar behaviour for l_s to that of basal faces, but displaced to somewhat higher temperatures as shown by the broken curve in fig. 5.7, then regions in which there is a strong asymmetry between migration conditions upon basal and prism faces are sharply delineated. Whilst the temperature domains for different growth habits can be plausibly matched in this way, the exact mechanism leading to axial or lateral growth is not clear.

The rate of growth of ice crystals from the vapour is of obvious meteorological significance in addition to its intrinsic interest. Generally speaking, however, ice crystal growth within clouds is limited by the diffusion of heat and of water molecules rather than by surface processes, and interest has centred on solution of the growth equations (5.19) and (5.20) or their more general forms allowing for motion of the growing crystal through its environment. The solution of these two equations for an environment with a supersaturation $\Sigma = (p - p_S)/p_S$ with respect to ice yields a rate of increase of crystal mass m of

$$dm/dt = 4\pi C X \Sigma, \tag{5.28}$$

where
$$X = D\rho_V \left[1 + \frac{DL^2\rho_V}{RT^2M_0\kappa}\right]^{-1}, \tag{5.29}$$

ρ_V and ρ_S are densities of vapour and solid respectively, M_0 is the molecular weight of water and κ the thermal conductivity of the vapour–air mixture (Mason 1953). Since C is proportional to a

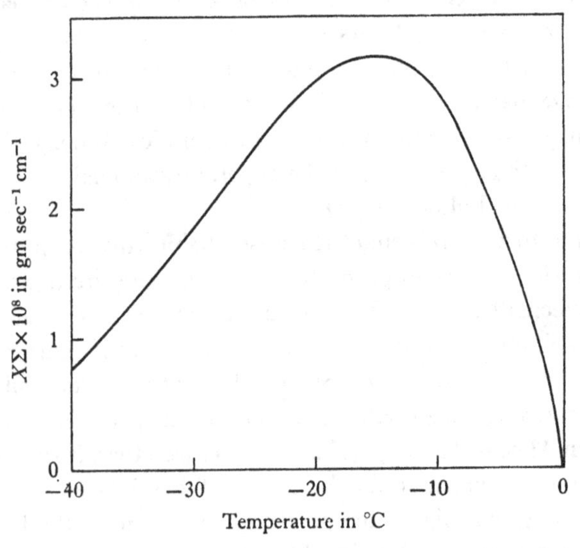

Fig. 5.8. The quantity $X\Sigma$ plotted as a function of temperature for growth of an ice crystal in an environment at water saturation.

characteristic dimension r of the crystal, (5.28) shows that r^2 increases linearly with time at a rate proportional to Σ. The quantity $X\Sigma$, for the meteorologically important case of growth from an environment at water saturation, has a maximum at about −15 °C, as shown in fig. 5.8.

When an ice crystal is growing upon a substrate the inhibiting effect of heat flow is much less severe and a situation may be reached where interface processes become important. This was the case with studies of the growth of ice crystals upon a metal plate, reported by Shaw & Mason (1955). For a crystal of axial lengths c

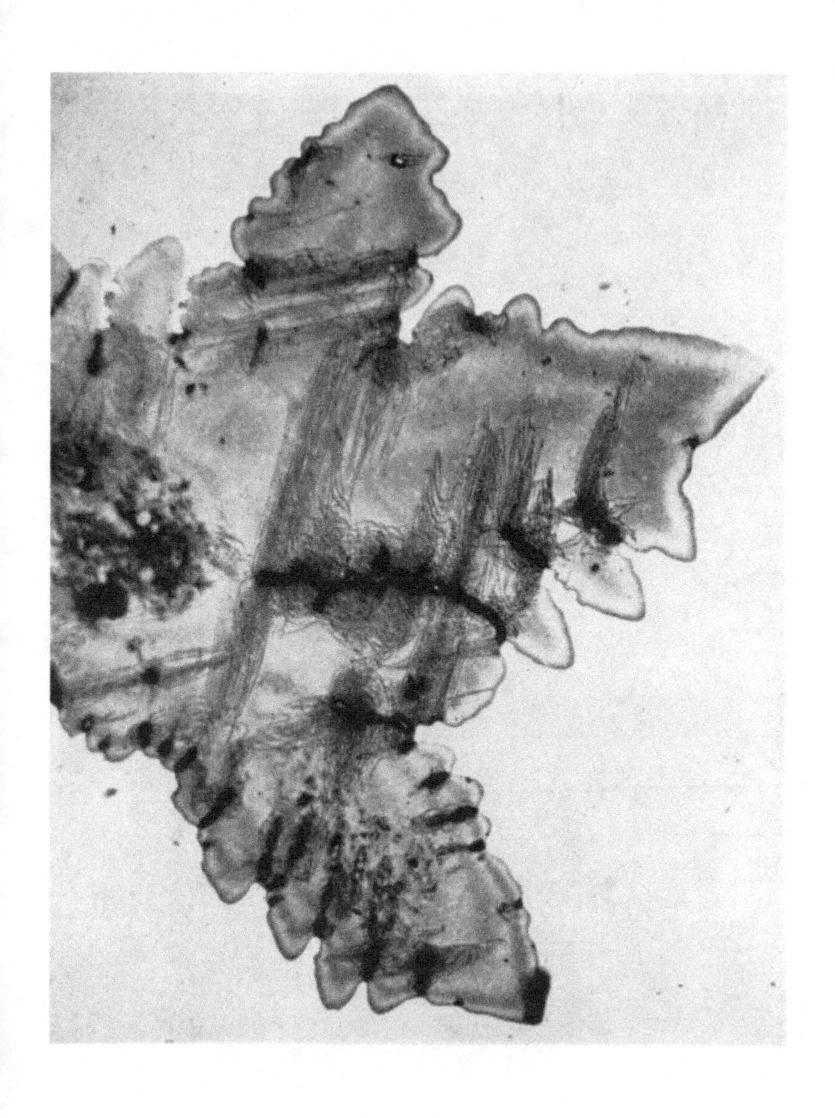

3 X-ray diffraction topograph of a portion of a tabular ice dendrite after gentle deformation. Dislocations appear as thin dark lines, and smoothly shaded regions of crystal are virtually perfect. The fraction of crystal shown is about 1 cm across (Webb & Hayes, 1967).

(*Facing p. 128*)

and a respectively they found both dc^2/dt and da^2/dt to be constant with time, which is a modified form of the simple behaviour discussed above. The limiting crystal habit, which varies considerably with temperature, is determined by the ratio of the square roots of these two quantities. They also found, however, that dc^2/dt and da^2/dt varied with the square of the supersaturation Σ for values of Σ from about $-0{\cdot}3$ (evaporation) to $+0{\cdot}4$ (growth), as would be expected from (5·11) for growth controlled by a dislocation mechanism at the interface, rather than the linear law (5.28) expected for diffusion-controlled growth. The actual magnitude of the growth velocity varied randomly from one crystal to the next.

CHAPTER 6

THERMAL PROPERTIES AND LATTICE DYNAMICS

In this chapter we shall discuss those properties of ice crystals which derive essentially from the thermal motions of water molecules within the crystal structure. In broad outline the theory describing these phenomena is simple and well known and leads to simple generalizations like the Debye theory of specific heats. However, because of the structure of the water molecule and, deriving from it, the structure of the ice crystal, such theories in their simple form represent only a first approximation to the observed behaviour. The coefficient of thermal expansion, for example, is negative at low temperatures and the specific heat is only poorly described by a Debye curve. It will be in tracing the reasons for some of these deviations from simple behaviour that most of our interest will lie.

6.1. Thermal expansion

The coefficient of thermal expansion can be determined in two different ways. The first, and most direct, is simply to take a single crystal of ice and measure its dimensions as a function of temperature. From the symmetry of the crystal the results can be expressed in terms of two linear expansion coefficients: $\alpha = l^{-1}(dl/dT)$ in directions parallel to the c-axis and to an a-axis respectively. Alternatively X-ray diffraction methods can be used to measure the c and a dimensions of the unit cell as a function of temperature. These two methods do not, in fact, measure exactly the same thing, since in a real crystal there will be an equilibrium concentration of vacancies and interstitial molecules, and these concentrations will change with temperature. As we shall see, however, the experimental results for ice are not sufficiently accurate to enable a meaningful distinction to be drawn.

Figure 6.1 summarizes the best experimental data available. Most measurements have been made with macroscopic single crystals and recent work has been in close agreement with the early results of Jakob & Erk (1928), who used polycrystalline samples. Hamblin (see Powell, 1958) was the only person to find a measurable difference between α_c and α_a for macroscopic measurements and in his case α_c only exceeded α_a by 1×10^{-6} deg^{-1}, which represents

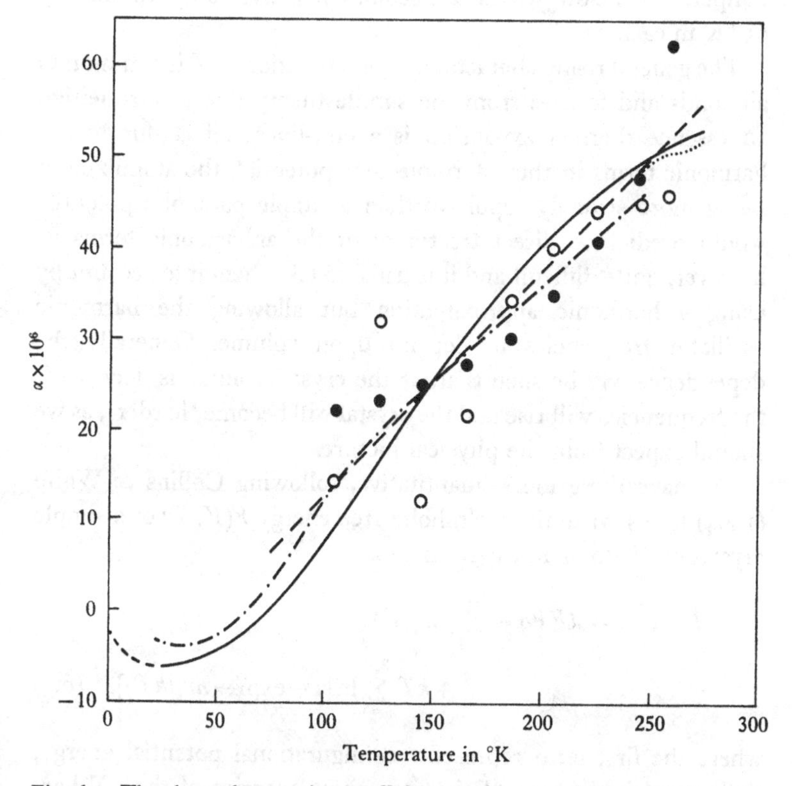

Fig. 6.1. The thermal expansion coefficient α of ice: as determined for polycrystalline samples by Jakob & Erk (1928), —; for single crystals by Hamblin (Powell, 1958), – –; Butkovich (1959), · · · · ·; and Dantl (1962), –·–; and from X-ray measurements by LaPlaca & Post (1960), ● α_c, ○ α_a.

about 2 per cent near 0 °C. The X-ray measurements of LaPlaca & Post (1960) referred to in chapter 2 are the only set complete enough to yield expansion coefficients. The values found for α_c and α_a are not significantly different except perhaps near the melting point, where α_c is rather larger than α_a, and near 130 °K where there seems to be some sort of anomaly in α_a. Since the reality of these effects requires confirmation we will not discuss them further here.

Dantl (1962) has also measured the expansion coefficient of D_2O ice over the range 20–270 °K and finds only very minor deviations from the behaviour for H_2O ice. In particular, the temperature below which α becomes negative is approximately 65 °K in each case.

The general rising character of α as a function of T is common to all solids and follows from the simple theory due to Gruneisen. In essence thermal expansion is a non-linear effect due to anharmonic terms in the intermolecular potential, the atomic cores being more strongly repulsive than a simple parabolic potential would predict. A direct treatment of the anharmonic terms is, however, quite difficult and it is usual to take them into account by using a harmonic approximation but allowing the harmonic oscillator frequencies to depend upon volume. Generally the dependence will be such that, as the crystal volume is decreased, the frequencies will rise and the crystal will become 'harder', as we should expect from the physical picture.

To make these ideas quantitative, following Collins & White (1964) let us write the Helmholtz free energy $F(V, T)$ of a simple crystal of N atoms in the form

$$F(V,T) = U(V) + \sum_{i=1}^{3N} \tfrac{1}{2} h\nu_i(V) + kT \sum_{i=1}^{3N} \ln[1 - \exp(-h\nu_i/kT)], \quad (6.1)$$

where the first term represents configurational potential energy, the second is the sum of the zero-point energies of the $3N$ harmonic oscillators of frequencies ν_i describing the vibrations of the lattice, and the third is the thermal energy of these oscillators. From the definition of the coefficient of volume expansion

$$\beta = \frac{1}{V}\left(\frac{\partial V}{\partial T}\right)_p \quad (6.2)$$

and the isothermal compressibility

$$\chi = -\frac{1}{V}\left(\frac{\partial V}{\partial p}\right)_T \quad (6.3)$$

it follows by the methods of thermodynamics that

$$\beta/\chi = (\partial S/\partial V)_T = -\partial^2 F/\partial V \partial T. \tag{6.4}$$

Thus, from (6.1) and the thermodynamic result

$$C_V = -T(\partial^2 F/\partial T^2)_V \tag{6.5}$$

we have the Gruneisen relation

$$\beta = 3\alpha = \gamma C_V \chi / V, \tag{6.6}$$

where $\gamma(T)$ is defined by

$$\gamma(T) = \sum_{i=1}^{3N} \gamma_i C_i \Big/ \sum_{i=1}^{3N} C_i, \tag{6.7}$$

C_i is the contribution to the specific heat made by the ith normal mode and

$$\gamma_i = -\left(\frac{\partial \ln \nu_i}{\partial \ln V}\right)_T. \tag{6.8}$$

In the simplest case all the γ_i are the same, $\gamma(T)$ is constant and, from (6.6), α is simply proportional to C_V and vanishes like T^3 near 0 °K, rising to a constant value when T is well above the Debye temperature. More complex behaviour can be ascribed to the existence of different γ_i values for different lattice modes.

Since for ice, from fig. 6.1, α is negative in the region between 0 °K and about 65 °K, this implies that there are some lattice modes whose contribution to the specific heat in this range is large and for which γ_i is negative. The analysis of the lattice vibrational spectrum of ice made by Leadbetter (1965), to which we shall return later in this chapter, suggests that there is a peak in the density of states near an (optical) wave number of 46 cm^{-1} and that this peak contains about 7 per cent of the $3N$ modes for translation of molecules as a whole. The nature of the molecular motion in the modes of this peak is almost certainly due to transverse waves in the lattice, as we shall see in the next section. If the γ_i for these modes are negative then this provides the mechanism leading to negative α at very low temperatures. The normal longitudinal vibrations of the lattice would not be expected to have negative γ_i and the librational modes (hindered rotations) have too high a frequency ($\sim$600 cm^{-1}) to contribute appreciable C_i at these temperatures. Since the frequencies of translational modes in D_2O ice differ from those in H_2O ice only by a factor close to $(18/20)^{\frac{1}{2}}$, its behaviour would be expected to be very similar.

It is interesting to note in passing that many other tetrahedrally bonded crystals have a similar negative expansion coefficient at low temperatures, for example germanium (< 48 °K), silicon (< 120 °K) and diamond (< 90 °K), and the same applies to materials like vitreous silica (Collins & White, 1964). This behaviour is thus apparently a feature of this type of bonding structure, rather than being a peculiarity of ice itself.

6.2. Heat capacity and lattice vibration spectrum

The heat capacity at constant pressure, C_p, has been measured for H_2O ice over the range 15–273 °K by Giaque & Stout (1936) and over the range 2–27 °K by Flubacher *et al.* (1960), the agreement in the range of overlap being very good. Similar measurements for D_2O ice over the range 15–277 °K have been reported by Long & Kemp (1936). The results of these measurements are summarized in fig. 6.2.

Before trying to interpret these results we need to express them in terms of C_v using the usual thermodynamic relation

$$C_p - C_v = \beta^2 VT/\chi. \tag{6.9}$$

From the thermal expansion data given in fig. 6.1 and the elastic moduli, which we shall discuss in chapter 8 and which indicate that χ ranges from about 11×10^{-12} dyne^{-1} cm^2 near 0 °K to about 13×10^{-12} dyne^{-1} cm^2 at the melting point, it is found that the correction to C_v makes very little difference to fig. 6.2. The value for C_v is about 3 per cent less than C_p near the melting point and the relative correction is much smaller at lower temperatures. For all but the most accurate work, therefore, we may neglect the distinction.

From fig. 6.2(*a*) it is immediately apparent that the specific heat of ice differs considerably from that predicted by a simple Debye treatment. Fig. 6.2(*b*) shows the deviation from the T^3 behaviour expected at very low temperatures and there is no sign of a flattening of the specific heat at temperatures up to the melting point. The limiting value of C_p/T^3 as T approaches 0 °K does,

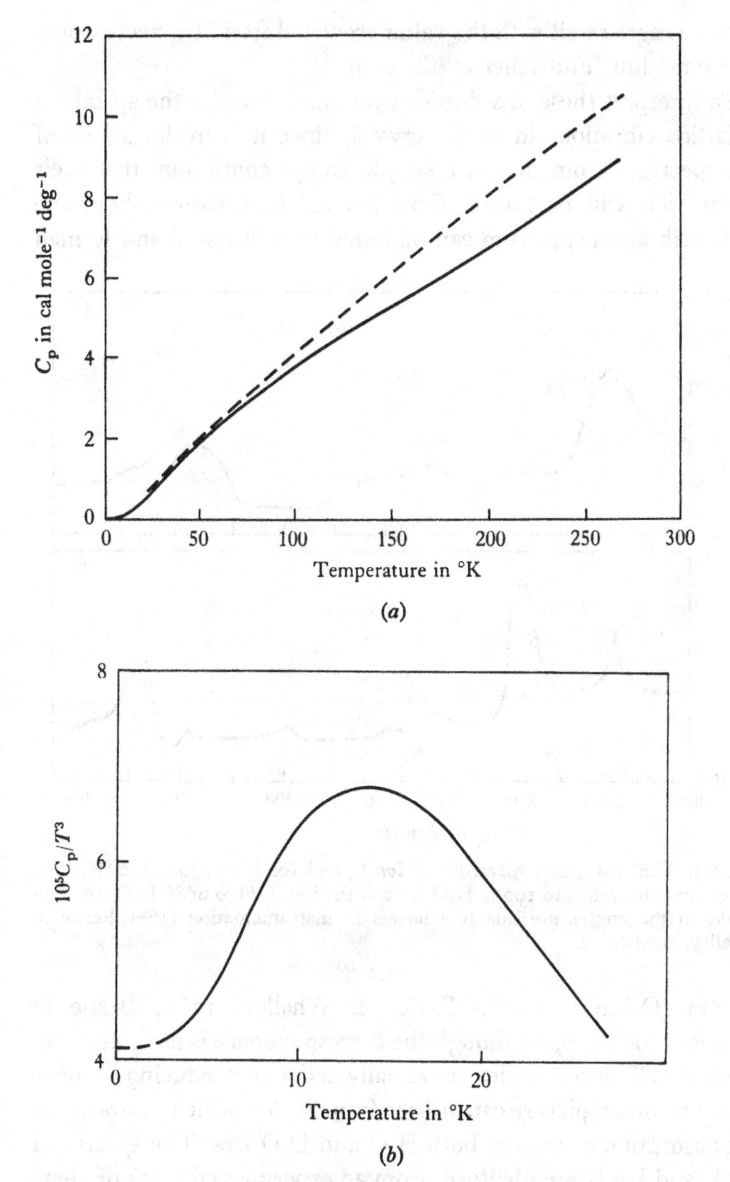

Fig. 6.2. (*a*) The heat capacity of H_2O ice as determined by Giaque & Stout (1936) and by Flubacher *et al.* (1960), and (broken curve) of D_2O ice as determined by Long & Kemp (1936). (*b*) A curve for C_p/T^3 from the data of Flubacher *et al.* for H_2O.

however, agree well with the value predicted from the macroscopic elastic moduli (Flubacher *et al.* 1960).

To interpret these observations we must consider the spectrum of lattice vibrations in an ice crystal, since it is in deviations of this spectrum from that of a simple Debye continuum that their explanation will be found. Experimental information about the lattice vibration spectrum can be found from infrared and Raman

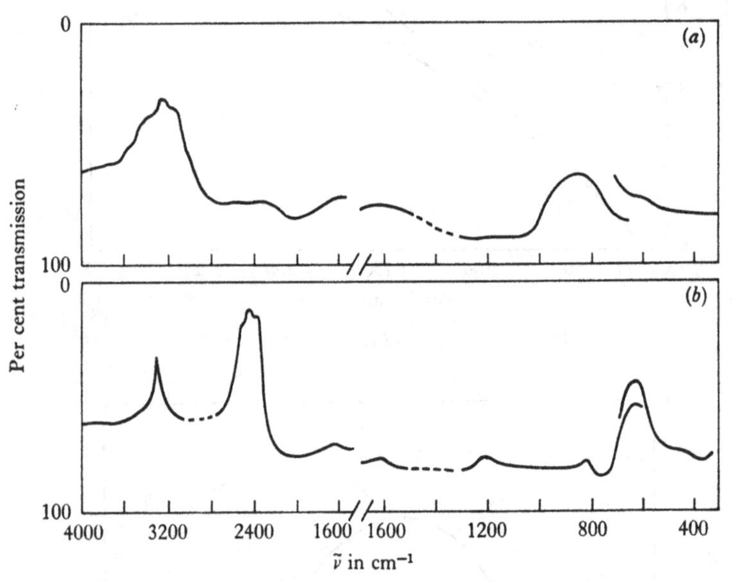

Fig. 6.3. The absorption spectrum of Ice I_h and Ice I_c at about 110 °K. The upper curve (*a*) refers to 100 % H_2O ice and the lower (*b*) to 95 % D_2O ice. The breaks in the curves are due to changes in instrumentation (after Bertie & Whalley, 1964*a*).

spectra (Ockman, 1958; Taylor & Whalley, 1964; Bertie & Whalley, 1964*a*, 1967) though the correspondence is not complete, since not all lattice modes are equally active in producing absorption. The broad picture can be seen from fig. 6.3, which shows infrared absorption spectra of both H_2O and D_2O ices. The spectra of Ice I_c and Ice I_h are identical, from an experimental point of view.

Three different types of modes contribute to the lattice spectrum and to infrared absorption. In the first place there are intramolecular modes corresponding to combinations of the three fundamental modes for an isolated water molecule which we discussed in chapter 1, their frequencies being only slightly changed by the effects of hydrogen bonding. In H_2O ice the $\tilde{\nu}_1$ and $\tilde{\nu}_3$ modes contribute to the complex peak near 3200 cm^{-1} which is shifted to near 2400 cm^{-1} in D_2O ice. This frequency shift, corresponding closely to a factor $2^{\frac{1}{2}}$, serves to check the identification. Details of these fundamental frequencies are given in table 6.1. Combinations of these fundamentals with each other and with other lattice modes give further peaks at higher frequencies, an example of which is the peak near 3200 cm^{-1} in D_2O ice.

TABLE 6.1. *General features of the infrared and Raman spectra of Ice I (Ockman, 1958; Taylor & Whalley, 1964; Bertie & Whalley, 1964a, 1967) for H_2O and D_2O.*

Intramolecular	$\tilde{\nu}_1$ or $\tilde{\nu}_3$	3150–3380 cm^{-1}	2330–2420 cm^{-1}
	$\tilde{\nu}_2$	1650 ± 30	1210 ± 10
Libration		500–1050	350–750
Translation		65, 164, 229	~60, 156, 222

The other vibrational modes of the crystal may be called intermolecular, since they involve interactions between neighbouring molecules. The intermolecular modes may be of two types, either translational, in which case the frequency shift in going from H_2O ice to D_2O ice will be close to $(18/20)^{\frac{1}{2}}$, or rotational, in which case the shift will be $(1/2)^{\frac{1}{2}}$. The rotational modes, which are often called librations after the astronomical term, contribute the broad peak extending from 500 to 1050 cm^{-1} in H_2O ice and from 350 to 750 cm^{-1} in D_2O ice. The translational modes, which we shall analyse in more detail, give a set of peaks below about 500 cm^{-1}, the most prominent of which are listed in table 6.1.

The translational modes, which are of course the only modes which can exist in simple monatomic crystals, can be treated in more detail. This is appropriate since we saw in the previous section that it is just these vibrations which are responsible for the negative expansion coefficient of ice at low temperatures. We said before that there is no exact correspondence between lattice modes and infrared spectra because of the different coupling of different modes to the electromagnetic field. With translational vibrations of an orientationally disordered crystal-like ice, however, the situation becomes rather different and Whalley & Bertie (1967) have shown that modes which are otherwise optically inactive can interact with the field on account of the disorder, the interaction being approximately proportional to the square of the frequency ν. This suggests that if, instead of optical density, we plot this quantity divided by ν^2, we should have a semi-quantitative representation of the

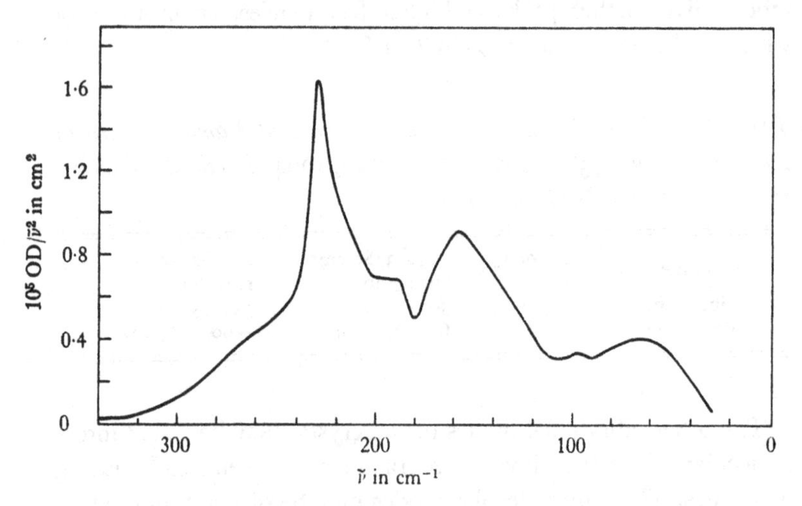

Fig. 6.4. The function (optical density)/(frequency)^2 plotted against wave number $\tilde{\nu}$ for H_2O ice to give a semiquantitative representation of the spectrum of vibrational states (Bertie & Whalley, 1967).

density of states of the vibrational spectrum. This procedure was followed by Bertie & Whalley (1967) with the result shown in fig. 6.4.

Since the spectra of Ice I_h and Ice I_c are identical, it makes sense to analyse the vibration spectrum indicated in fig. 6.4 in terms of the simpler diamond cubic structure of Ice I_c. In fact, since the molecules move as rigid units for the translational modes, the analysis should be very similar to that for the lattice vibrations of diamond, silicon or germanium. All these crystals have two atoms per unit cell (two molecules in the case of Ice I_c) and the vibrational spectrum has two branches: an acoustic branch, in which the two atoms move essentially in phase, and an optical branch, in which their motion is antiphase (Ziman, 1960, chapter 1). In addition, each of these branches is split into one longitudinal and two transverse sub-branches. The transverse acoustic branch lies below the longitudinal acoustic branch because the shear modulus is less than the bulk modulus for all ordinary materials. Because of the crystal symmetry and periodicity the dependence of frequency upon wave vector for each of these branches can be

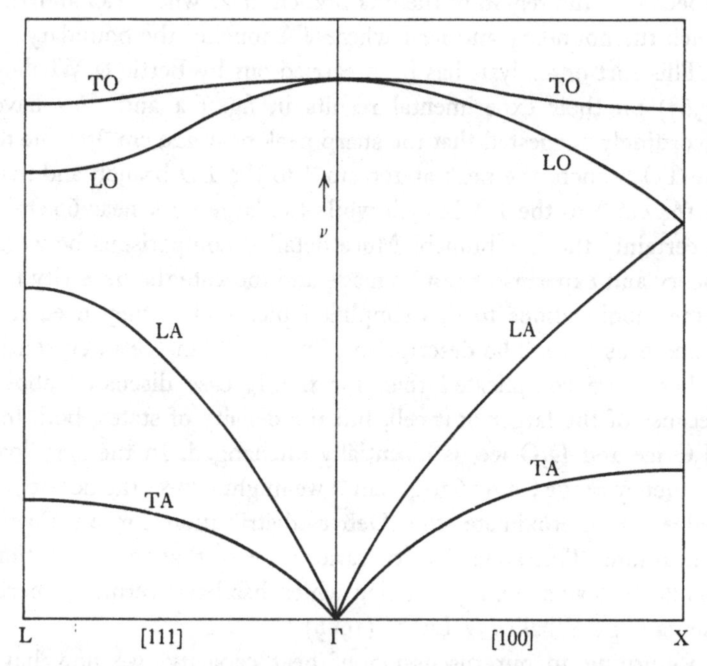

Fig. 6.5. Dispersion curves giving frequency ν in terms of wave vector $\mathbf{k}$ in the [100] and [111] directions for a crystal with the diamond structure (L = longitudinal, T = transverse, A = acoustic, O = optical).

plotted in the first Brillouin zone of the reciprocal lattice and the result, for two particular propagation directions in which the transverse modes are degenerate, has the form shown in fig. 6.5. Γ is the zone centre, X the edge of the zone in a [100] direction and L the edge of the zone in a [111] direction. The corresponding curves for ice will almost certainly differ in detail from those of fig. 6.5 but the general sequence and behaviour must be similar.

Now the density of states in wave-number space ($\mathbf{k}$-space) is constant so that in the range (k, dk) the number of states is proportional to k^2. Thus, when we try to determine the density of states as a function of frequency, the largest contribution comes from regions where the curves in fig. 6.5 are nearly horizontal, and particularly from such regions near the edge of the zone where k is large. High densities of vibrational states can therefore be expected in the region of the TO branch, near where LO and LA touch the boundary and near where TA touches the boundary.

This sort of analysis has been carried out by Bertie & Whalley (1967) on their experimental results in fig. 6.4 and they have accordingly suggested that the sharp peak near 229 cm^{-1} is due to the TO branch, the peak at 190 cm^{-1} to the LO branch and that at 164 cm^{-1} to the LA branch while the large peak near 60 cm^{-1} is certainly the TA branch. More detailed comparisons between theory and experiment can be made and indicate the necessity for some modifications to this simplified picture but they need not concern us here. The description of modes in an Ice I_h crystal is rather more complicated than for the I_c case discussed above because of the larger unit cell, but the density of states, both for H_2O ice and D_2O ice, is essentially unchanged. In the very low frequency range from 16 to 40 cm^{-1} we might expect the density of states to approximate the Debye distribution for an elastic continuum. This state density varies as ν^2 so that the absorption should behave as ν^4, a prediction which has been verified experimentally by Whalley & Labbé (1969).

Returning to our discussion of heat capacity, we now have enough information to see what determines the behaviour shown in fig. 6.2. The intramolecular vibrations, all having wave numbers in excess of 1600 cm^{-1} for H_2O ice, have characteristic temperatures (defined by $h\nu = k\Theta$) above 2300 °K and so contribute very little to the specific heat. The librational modes have a peak near 600 cm^{-1} (Θ = 860 °K) so that their contribution is considerable at temperatures approaching the melting point. The translational modes form a continuum and so contribute largely to the specific heat at all temperatures. The large peak near 60 cm^{-1} (Θ = 86 °K) means that the behaviour will differ from the simple Debye form at temperatures of only a few tens of degrees.

There are two approaches to the more detailed study of the

specific heat curves. Either we can assume a density-of-states distribution on the basis of experimental evidence and, by adjusting a few parameters, try to bring the computed heat capacity to agreement with experiment, or we can start from the experimental heat capacity curve and from it deduce certain properties of the density of states. The first path was followed by Bjerrum (1951), who represented the librational oscillations by a Debye spectrum of weight 3 and characteristic temperature corresponding to 800 cm^{-1}, and the translational part of the spectrum by another Debye function of weight 3 and characteristic temperature corresponding to 210 cm^{-1}, these frequencies being chosen to agree with infrared evidence available at the time. Agreement between the calculated and experimental heat capacities was moderately good ($\pm$ 10 per cent) above about 50 °K but the calculated values were markedly low below this temperature.

A rather different sort of calculation was made by Blue (1954), who observed that the main difference between the vibrational spectra of H_2O ice and D_2O ice arises from the frequency shift of a factor $(\frac{1}{2})^{\frac{1}{2}}$ in the librational modes. In the region between 100 and 200 °K the difference between the two C_p values can, he found, be fitted to within 3 per cent by the difference between two Einstein functions located at Θ = 945 °K ($\tilde{\nu}$ = 660 cm^{-1}) for H_2O ice and appropriately shifted for D_2O ice. Still better agreement over the full temperature range 0 to 273 °K was obtained by representing the librational modes by an Einstein function with Θ = 1040 °K ($\tilde{\nu}$ = 723 cm^{-1}) and the translational modes by a Debye function with characteristic temperature 315 °K for H_2O ice, using appropriately shifted frequencies for D_2O and small corrections for intramolecular modes.

These calculations are instructive and serve to confirm the theoretical treatment we have outlined but they do not give any really new information. Leadbetter (1965), however, by proceeding in the opposite direction from the experimental heat capacity, was able to deduce a number of features of the vibrational spectrum. Broadly speaking the analysis goes as follows. At very low temperatures the heat capacity of an insulator can be expanded as

$$C_v = aT^3+bT^5+cT^7+\dots \tag{6.10}$$

and the expansion coefficients $a, b, c, \ldots$ are directly related to the coefficients $a', b', c', \ldots$ in the expansion for the density of vibrational states at low temperatures

$$g(\nu) = 3N(a'\nu^2 + b'\nu^4 + c'\nu^6 + \ldots). \tag{6.11}$$

The values found for the coefficients were

$$\left.\begin{aligned} a' &= (7{\cdot}69 \pm 0{\cdot}23) \times 10^{-7} (\mathrm{cm}^{-1})^{-3}, \\ b' &= (3{\cdot}5 \pm 1{\cdot}7) \times 10^{-10} (\mathrm{cm}^{-1})^{-5}, \\ c' &= (5 \pm 4) \times 10^{-14} (\mathrm{cm}^{-1})^{-7}. \end{aligned}\right\} \tag{6.12}$$

At somewhat higher temperatures the spectrum can be analysed in terms of a monochromatic peak superimposed on a Debye spectrum and the conclusion is that such a peak is located at 48 cm^{-1} for H_2O ice and contains about 7 per cent of the $3N$ translational modes. This agrees quite well with the spectroscopic evidence discussed earlier and with the fact that the TA branch contains a total of N vibrational modes.

Finally, from a more extended temperature range, the moments $\langle \nu^n \rangle$ of the vibration spectrum can be found and, since the low-temperature expressions consider only translational modes, it is possible to find separate moments for the contributions of the librations. These moments give general information about the shape of the spectrum, and those for librations only can be interpreted quite simply since $\langle \nu_{\mathrm{lib}}^n \rangle^{1/n}$ turns out to depend only slightly on n, indicating a narrow peak. For H_2O ice the librational band is centred on 650 cm^{-1} and has a half-width of about 200 cm^{-1}, while the corresponding figures for D_2O ice are 520 cm^{-1} and 160 cm^{-1}. These conclusions are again in excellent agreement with those of infrared studies summarized in table 6.1.

All these studies thus agree upon the general features of the vibrational spectrum of ice, though there is room for refinement of our detailed knowledge.

Whilst it is clear that any simple Debye-like treatment of lattice vibrations in ice is inadequate, it is useful for some semi-quantitative purposes to describe the state of the crystal in terms of an equivalent Debye characteristic temperature which will itself vary with the temperature of the crystal. If intramolecular motions and librations are neglected, then the crystal has $3N$ degrees of freedom, where N is the number of molecules, and the effective Debye temperature derived from specific heat data is 220 °K at $T = 0$ °K, falls to a minimum of 180 °K at $T = 16$ °K, rises to 320 °K at $T = 100$ °K and then falls to 100 °K at T = 170 °K as the librational modes become active (Leadbetter, 1965). These results are roughly consistent with the Debye temperature of 224 °K used by Zajac (1958) in interpretation of X-ray diffraction data but clearly no close agreement is to be expected. More meaningful are the actual r.m.s. vibrational amplitudes of atoms in the different modes. These are discussed and compared with experiment by Leadbetter and a selection of values is given in table 6.2. The largest contribution at all temperatures comes from translational oscillations of whole molecules. The r.m.s. librational amplitude for hydrogen ranges only from 0·12 to 0·17 Å and for deuterium from 0·087 to 0·094 Å, while the intramolecular OH and OD stretching vibrations have essentially only their zero-point amplitudes of 0·041 and 0·035 Å respectively.

TABLE 6.2. *R.m.s. vibration amplitudes (Å) for atoms in* H_2O *and* D_2O *(interpolated from Leadbetter,* 1965)

Temperature (°K)	O	H	D
0	0·092	0·150	0·129
100	0·132	0·178	0·164
200	0·185	0·221	0·206
273	0·215	0·248	0·236

6.3. Thermal conductivity

The thermal conductivity of ice is the final topic which we shall take up in the present chapter. The experimental situation here is not very satisfactory and, as shown in fig. 6.6, there are large temperature ranges where no measurements have been made. The results of Jakob & Erk (1929) and of Ratcliffe (1962) are in moderately good agreement between about 100 °K and the melting point and show an approximate T^{-1} dependence which, as we shall see presently, is to be expected. This inverse dependence might reasonably be extrapolated to 20 °K to give good agreement with the results of Dean & Timmerhaus (1963) at this temperature but there is a suggestion in the experimental results that there is some sort of hump in the curve as shown by the broken interpolation, and indeed this sort of behaviour is to be expected on theoretical grounds.

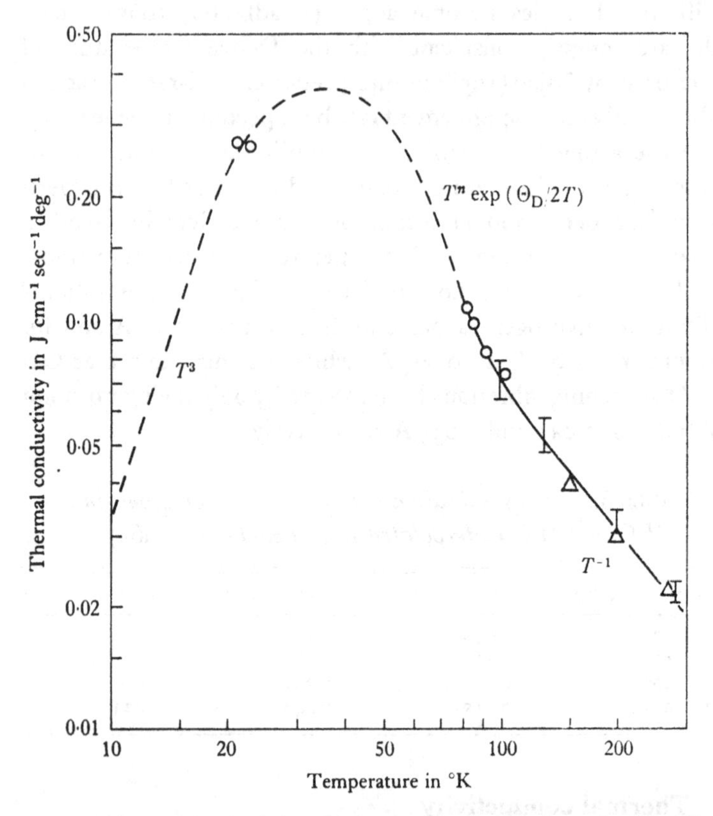

Fig. 6.6. Thermal conductivity of polycrystalline ice: as determined by Jakob & Erk (1929), △; Ratcliffe (1962), I; and Dean & Timmerhaus (1963), O. The behaviour suggested by theory is shown as a broken interpolation curve.

The measurements shown in fig. 6.6 are for polycrystalline specimens of ordinary ice; measurements on D_2O ice near 20 °K and 100 °K by Dean & Timmerhaus (1963) show conductivities about 10 per cent lower. The few measurements which have been made on single crystals of ordinary ice (Powell, 1958) suggest that κ is about 5 per cent greater in a direction parallel to the c-axis than normal to this direction.

No detailed theoretical discussion of thermal conduction in ice has yet been given but the general outline of the theory is the same as that for other electrically non-conducting crystals (Ziman, 1960, chapter 8; Kittel 1966, chapter 6). The thermal conductivity can be thought of quite generally as due to diffusion of phonons down the temperature gradient, with mean speed $\bar{v}$ and mean free path Λ, giving for the thermal conductivity κ the simple relation

$$\kappa = \tfrac{1}{3}C_{\mathrm{v}}\bar{v}\Lambda \tag{6.13}$$

where C_{v} is the heat capacity per unit volume. Taking $\bar{v}$ equal to the speed of sound in ice, which is about $2{\cdot}2 \times 10^5$ cm sec^{-1} at -16 °K (Flubacher *et al.* 1960), we find mean free paths ranging from about 27 Å at 100 °K to 15 Å at the melting point.

The mean free path Λ may be determined by many different scattering mechanisms but the dominant one at temperatures not too close to 0 °K is phonon–phonon scattering, the coupling taking place through the anharmonicity of the lattice vibrations. There are two possible types of phonon–phonon scattering processes: normal processes in which total phonon wave vector is conserved, and umklapp processes in which the total wave vector after collision differs from that before collision by a vector of the reciprocal lattice. Since normal processes do not affect the total phonon momentum or energy, they do not contribute to thermal resistance and only umklapp processes need be considered. For an umklapp process to occur between two phonons of wave vectors $\mathbf{q}$ and $\mathbf{q}'$ we must have a relation of the form

$$\mathbf{q}+\mathbf{q}' = \mathbf{q}''+\mathbf{K} \tag{6.14}$$

where $\mathbf{q}''$ is the wave vector of the resulting phonon and $\mathbf{K}$ is a reciprocal lattice vector. If (6.14) is to be satisfied and phonon energy conserved at the same time, then at least one of $\mathbf{q}$, $\mathbf{q}'$ must be comparable to $\mathbf{K}$ in magnitude.

Above the Debye temperature Θ_{D}, for a simple solid, phonons with large wave vector are common and increase in number in proportion to T, while C_{v} is nearly constant. We thus expect κ to vary as T^{-1}. This is the observed behaviour for ice above 100 °K and, though the processes are more complex than this, we might reasonably associate the principal scattering mechanism in this temperature region with the transverse acoustic phonons which, as we have seen, have a characteristic temperature of rather less than 90 °K.

Below Θ_D, for a simple solid, the excitation of phonons of sufficiently large wave vector for umklapp scattering requires energy greater than about $\frac{1}{2}k\Theta_D$, so that their concentration varies roughly as $\exp(-\Theta_D/2T)$ and, supposing C_v to behave as T^n in this range,

$$\kappa \propto T^n \exp(\Theta_D/2T), \tag{6.15}$$

which rises sharply as the temperature falls. There is a suggestion of such a rise in the ice results just below 100 °K.

For still lower temperatures the mean free path Λ is limited by scattering from impurities and imperfections, rather than by phonon–phonon interactions, and becomes independent of temperature. The specific heat C_v in this region tends towards a T^3 dependence, so κ itself varies as T^3. This behaviour is suggested by the extrapolation to low temperatures in fig. 6.6. The complete temperature dependence of κ for ice obviously requires further investigation, but this sort of general behaviour has been observed in materials like alumina (Ziman, 1960, p. 292), diamond and germanium (Kittel, 1966, pp. 192–3) over comparable temperature ranges.

CHAPTER 7

POINT DEFECTS

So far in this book we have been concerned almost entirely with the properties of perfect crystals of ice. The properties discussed were, in fact, not structure-sensitive and both the theory and necessarily the measurements apply equally to real crystals. In the remaining chapters, however, we shall examine attributes of ice which do depend sensitively upon the perfection of the particular crystal involved. There are many imperfections which occur in real crystals—surfaces, impurities, dislocations, vacancies and so on —with which we shall necessarily be concerned but in the present chapter we focus attention on the various possible kinds of point defects.

The structure of a perfect ice crystal is, as we have seen, of a statistical kind in which many different configurations are allowed provided they satisfy the three rules: (i) each lattice position is occupied by a water molecule tetrahedrally bonded to its four nearest neighbours; (ii) water molecules are intact so that there are just two protons near each oxygen; (iii) there is just one proton on each bond. Violation of the first rule leads to a vacancy, an interstitial or an impurity atom, while violation of the second or third gives rather more subtle defects peculiar to the ice structure. The very existence of these rules implies an energy penalty for their violation but, in any real crystal at a finite temperature, there will be a Boltzmann probability for finding such exceptions.

7.1. Ion states

A pair of ion states H_3O^+ and OH^- is formed, at least from a formal point of view, when a proton jumps from the normal end of a bond to a position near the other end as shown in fig. 7.1 *a*, thus violating rule (ii). According to the semi-empirical calculations of Lippincott & Schroeder (1955) plotted in fig. 2.6 there is a subsidiary potential minimum for a proton at this position on the bond, though the simplified quantum-mechanical treatment of Weissmann &

Cohan (1965*a*) plotted in fig. 2.5 does not show this. The two neighbouring ion states are thus at best metastable and may even be completely unstable towards recombination. There is, however, a finite possibility that, while the neighbouring dissociated state is in existence, a further proton jump may take place along another bond of one of the partners as in fig. 7.1*b*, so that the two ion states become separated. Once this has happened, the two defects are stable in the sense that, though there is an attractive electrostatic

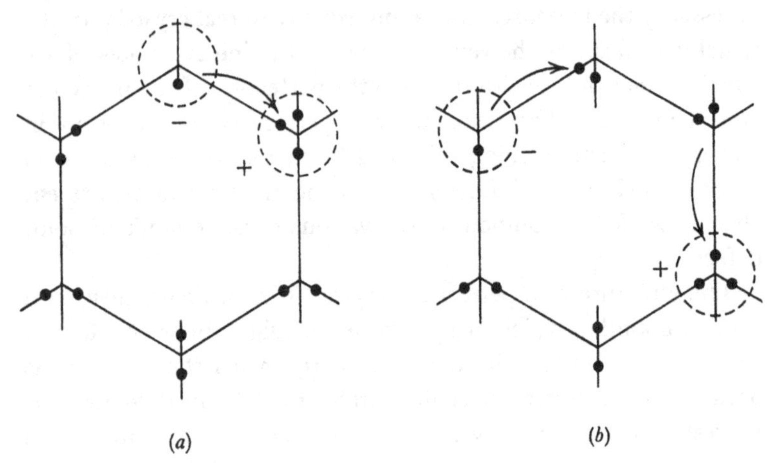

Fig. 7.1. (*a*) Formation of an ion-state pair, H_3O^+ and OH^-, by a proton jump along a bond. (*b*) Separation of these two ion states by further proton jumps.

force between them, they cannot immediately recombine. Further proton jumps may then lead either to such a recombination or to a diffusive separation of the two ion states.

Since all these processes take place in thermal equilibrium, the concentrations $n_\pm$ of ion states in pure ice are given by

$$n_+ = n_- = n_S \exp(-H^f_\pm/2kT), \qquad (7.1)$$

where $H^f_\pm$ is the enthalpy change involved in the formation of a single separated defect pair and, since the amount of dissociation is very small, n_S is very nearly the number of water molecules per unit volume in the crystal.

It is not easy to estimate the concentration of ion states to be expected in ice, but several possible approaches have been discussed in detail by Eigen & De Maeyer (1958). For liquid water the experimental problem is somewhat simplified because such a variety of possible solutes can be used and, as is well known, the concentration of H^+ or OH^- ions at 25 °C is $1{\cdot}0 \times 10^{-7}$ mole l^{-1} or 6×10^{13} ion pairs per cm^3, the dissociation enthalpy being 0·59 eV. Now the dissociation enthalpy $H^f_\pm$ is made up of two parts, that involved in proton transfer along the bond and a further part arising from separation of the two ion states against their electrostatic attraction. The first part is probably not very different in magnitude in ice and water, but there is an important distinction between the amount of energy required to separate the ions in the two cases. In liquid water the structure is rapidly fluctuating and molecular reorientation can occur very quickly and follow the motion of the ions. The dielectric constant, except for saturation effects very close to the ions, thus has approximately the macroscopic value of 80. In ice, on the other hand, molecular orientations can only occur as a result of the motion of defects and the characteristic relaxation time, as we shall see in chapter 9, is of the order of 10^{-4} sec. The reorientations are thus much less able to follow the proton motion and the dielectric constant is much less than the macroscopic value. The high-frequency value 3·2 is appropriate in this case, as will become clear from the discussion in chapter 9. Any calculation based on this sort of model is necessarily greatly oversimplified but the dissociation enthalpy estimated ranges from about 1 to 2 eV, depending on details of the assumptions (Bjerrum, 1951; Eigen & De Maeyer, 1958). Comparison with the $H^f_\pm$ value of 0·59 eV for liquid water and use of (7.1) then suggests that the concentration of ion states in ice should be less than that in water at the same temperature by a factor between 10^{-2} and 10^{-5}. As we shall see in a moment, the former factor is most nearly in agreement with experiment.

It is not a simple matter to measure the concentration of ion states in ice and several different methods have been used. The equilibrium concentrations $n_\pm$, given by (7.1), are not static but represent a balance between the rate at which defects are generated thermally in the otherwise perfect crystal and the rate at which they recombine with one another. The dissocation rate can be written for either sign of state

$$(dn_\pm/dt)_D = k_D n_S, \qquad (7.2)$$

where n_S is the concentration of water molecules, while recombination, involving as it does collisions between pairs of ion states of opposite sign, behaves like

$$(dn_{\pm}/dt)_R = -k_R n_{\pm}^2. \tag{7.3}$$

The condition that these two rates sum to zero then gives

$$n_{\pm}^2 = (k_D/k_R)n_S. \tag{7.4}$$

Both these rate constants k_D and k_R can, in principle, be found by disturbing the equilibrium and watching the recovery of the ion concentration, and this method has been used by Eigen *et al.* (1958, 1964). To find k_R they upset the ion concentration by applying to an ice crystal a short voltage pulse of large amplitude and then observed the recovery of the electrical conductivity, assumed to be directly proportional to $n_{\pm}$. The generation rate constant k_D, on the other hand, was evaluated by applying to an ice crystal an electric field strong enough to sweep out all the ion states before they could recombine. The field involved was about 20 kV cm^{-1} for an electrode separation of 0·1 mm. Under these conditions the current saturates and, from the saturation value, k_D can be found. In addition, from the absolute value of the field for which saturation occurs and from the shape of the current–voltage curve at lower fields, approximate values can be derived for the ionic mobilities μ_+ and μ_-. Their experimental results are summarized in table 7.1, together with some similar results for D_2O ice.

This is not, however, the only way in which these quantities can be determined, and Levi *et al.* (1963) have used a more orthodox approach in which the electrical conductivity of ice was studied as a function of the quantity of proton donor and proton acceptor impurity present for ice crystals cross-doped with ammonia and hydrogen fluoride. Their results, which differ slightly from those of Eigen *et al.*, are also shown in table 7.1.

Three things immediately stand out from this table: the concentration of ion states in ice is about two orders of magnitude less than in water (extrapolated to the same temperature); the positive ion states are very much more mobile than the negative ion states (the ratio is only 2:1 in water); and the absolute value of the

H_3O^+ mobility is very large compared with typical ionic mobilities in solids ($< 10^{-8}$ cm^2 V^{-1} s^{-1}), higher by two orders of magnitude than the H_3O^+ mobility in water, and indeed almost comparable to electron mobilities in metals and elemental semiconductors like germanium (10–10^4 cm^2 V^{-1} s^{-1}). We shall return to consider the mechanism of proton motion in more detail when discussing electrical properties; for the present let us give some attention to the structure of the ion states themselves.

TABLE 7.1. *Dissociation and mobility data for pure ice*

	H_2O Ice			D_2O Ice (Eigen *et al.* 1964)
	Eigen & De Maeyer (1958)	Eigen *et al.* (1964)	Levi *et al.* (1963)	
Temperature of measurement (°C)	−10	−10	−17	−10
Ion state concentration $n_\pm$ (cm^{-3})	$(2–9) \times 10^{10}$	8.4×10^{10}	—	2×10^{10}
Energy for ion-pair formation $H_\pm^f$ (eV)	—	0.96 ± 0.13	—	—
Rate constant for dissociation k_D (s^{-1})	3×10^{-9}	3.2×10^{-9}	—	2.7×10^{-11}
Rate constant for recombination k_R (cm^3 s^{-1})	10^{-7}–10^{-8}	—	—	—
Dissociation constant $K = k_D/k_R$ (cm^{-3})	0.01–0.3	—	0.1	—
H_3O^+ mobility μ_+ (cm^2 V^{-1} s^{-1})	0.1–0.5	0.075	0.02	0.012
OH^- mobility μ_- (cm^2 V^{-1} s^{-1})	$\leqslant 0.05$	—	0.01	—
Mobility ratio μ_+/μ_-	10–100	—	1 to 5	—

In the ion state H_3O^+ in ice, no distinction is possible between the three protons. They will lie approximately at three of the vertices of a regular tetrahedron with the oxygen nucleus at its centre and will also lie quite closely upon the bonds O–H...O. The ten electrons in the molecule will redistribute themselves into fairly close similarity with the molecular orbitals in the ammonia molecule, though attracted more closely to the extra charge on the oxygen nucleus and distorted by the electric field due to neighbouring molecules. It is a little more difficult to be sure of the structure of the OH^- state. The single proton will certainly lie upon a bond direction but the only tetrahedral hybridization of electron wave functions will be that impressed upon the molecule by interactions with its neighbours.

The only quantum-mechanical treatment of the ion states yet undertaken seems to be that of Weissmann & Cohan (1965*b*) who, following their earlier treatment of the normal hydrogen bond, simplify the problem of interaction with the tetrahedral environment by supposing that, for both defects, tetrahedral hybridiza-

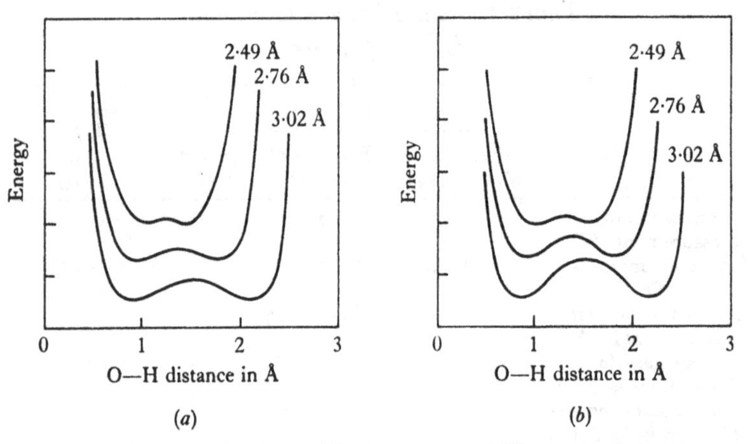

Fig. 7.2. Total energy for (*a*) positive ion states and (*b*) negative ion states as a function of O–H distance for several values of the O–O distance. The intervals on the energy scale are 5 eV and the zero of each curve is shifted vertically from the preceding one by exactly that amount (after Weissmann & Cohan, 1965*b*).

tion is complete. The quantity which they calculate is the energy of the system $H_2O\ldots HOH_2^+$ or $HOH\ldots OH^-$ for various O–O distances, as the position of the proton on the O...H–O bond is varied. This then gives an estimate of the structure of the ion state and of the height of the energy barrier to proton motion along the bond. The calculated curves are shown in fig. 7.2.

From these curves it is apparent that the energy barrier to proton motion in the transfer of an H_3O^+ state between neighbouring molecules is less than that for the transfer of an OH^- state, in agreement with the mobility data of table 7.1. This is not unexpected since, in the first case, we are transferring an excess proton between two basically neutral molecules while, in the second, the transfer takes place between two negatively charged OH^- units. The calculated energy barriers are shown in table 7.2, from which it can be seen that they depend sensitively upon the O–O distance. It is therefore possible that proton transfer may be coupled in some way with lattice vibration modes, though, as we shall see later, the proton motion is not a simple classical jump. The equilibrium O–O distance around the defect is not necessarily equal to the O–O distance in undisturbed ice, since the defect may distort the surrounding lattice; similarly the O–H distances may differ slightly from the normal values and figure 7.2 suggests that the O–H distance in H_3O^+ is slightly greater and that in OH^- slightly less than in the normal ice structure, as is indeed to be expected.

TABLE 7.2. *Calculated energy barriers to the motion of positive and negative ion states (Weissmann & Cohan, 1965b)*

O–O distance (Å)	Energy barrier (eV) H_3O^+	Energy barrier (eV) OH^-
2·49	0·19	0·62
2·76	0·98	1·90
3·02	2·20	3·50

7.2. Orientational defects

The concept of an orientational defect was put forward by Bjerrum (1951) to account for the dielectric properties of ice by introducing some mechanism, other than the motion of ion states, to allow the orientational relaxation of water molecules. These defects are produced, formally, by the rotation of a single molecule through $2\pi/3$ about one of its bond directions, leading to one doubly occupied bond (D-defect) and one unoccupied bond (L-defect) as shown in figure 7.3. The two oppositely signed defects can then diffuse further apart by successive rotations of neighbouring molecules. The molecular rotations involved can be pictured quite equivalently as involving an oblique jump of a single proton from one bond to another on the same molecule, the jump distance involved being about 1·65 Å.

Whilst the model of an L-defect as a bond direction with no proton on it is fairly satisfactory in the absence of discussion of the associated electronic structure, it is apparent that it is an over-simplification to consider a D-defect as an undistorted collinear structure with two protons lying on the bond. The two protons in such a structure, assuming the O–H distances to have their normal value, would be separated by only 0·74 Å, which is close to the internuclear distance in the molecule H_2 and much less than the distance to be expected for protons not bound together covalently.

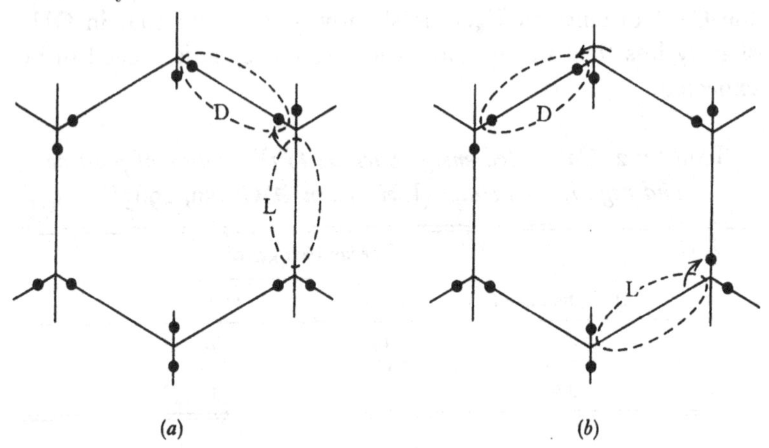

Fig. 7.3. (*a*) Formation of an orientational defect pair, D and L, by an oblique proton jump or, equivalently, by rotation of a water molecule by $2\pi/3$ about one of its bonds. (*b*) Separation of these two orientational defects by further oblique proton jumps.

Bjerrum recognized this difficulty when postulating the existence of these defects, though not actually distinguishing between the effect for L- and D-defects because of his regular tetrahedral point-charge model for the water molecule. He pointed out that one would expect the defect to distort the surrounding structure by increase of the O–O distance along the bond and by a small rotation so that the O–H...H–O configuration is no longer collinear. With Bjerrum's simple molecular model the energy of bonding is $-13\cdot6$ kcal mole^{-1} or $-0\cdot3$ eV per bond. At a defect site the electrostatic energy is reversed so that the formation energy of a defect is twice the bond energy or 0·6 eV. This energy is reduced by lattice distortion to an estimated 0·45 eV for either L- or D-defects.

Whilst this energy is very reasonable and allows the existence of a considerable concentration of defects in thermal equilibrium, it is evident that Bjerrum's symmetrical point-charge model for the water molecule is a very poor approximation when short-range interactions are considered and a better calculation is desirable. One such improved calculation was made by Cohan *et al.* (1962), again using a point-charge model but this time with seven individual charges: $+6e$, $+e$ and $+e$ at the three nuclei and four charges of $2e$ located respectively at the centroid of the charge distributions of each pair of bonding and lone-pair electrons, as evaluated for a combination of Duncan and Pople orbitals and tetrahedral hybrids. For this model, giving a bond energy of about $-0{\cdot}31$ eV in excellent agreement with the experimental value, the L-defect energy is 0·14 eV, giving an energy of formation of 0·45 eV. A collinear D-defect, in contrast, has a formation energy of about 1·4 eV, though this is reduced to roughly 0·5 eV if the two molecules involved each rotate by about 20° to make the structure non-collinear.

A physically more realistic calculation using a semi-empirical potential for the interaction between hydrogen atoms led Dunitz (1963) to an energy of about 3 eV for the collinear D-defect, though Eisenberg & Coulson (1963) showed that elastic relaxation of surrounding molecules, by separation and rotation, can reduce this energy to only 0·25 eV, which is quite close to the estimate for an L-defect. Dunitz had proposed that a D-defect should be effectively shared between two bonds by a rotation of one of the molecules normally involved through an angle of $\pi/3$, creating what he termed an X-defect, but this structure seems less likely than the elastically relaxed form of the simple D-defect.

In summary then, we see that theoretical estimates which take account of the elastic relaxation of the surrounding lattice put the formation energy of a pair of separated L- and D-defects at rather less than 1 eV. Experimental information from dielectric relaxation studies, which we shall discuss in chapter 9, gives a value of 0·68 eV which, in view of the uncertainties involved in the theory, represents reasonably satisfactory agreement.

This information allows us to make tentative estimates of the concentration of orientational defects in pure ice, using an equation like (7.1), and of their mobility. It is clear from the energies involved that they should be much more numerous in pure ice than are the ion states. The energy barrier to proton motion is comparable in height to that for ion states but twice as wide, so that it is possible, and indeed turns out to be the case, that the anomalously high mobility of ionic states does not extend to orientational defects. Experimental information, derived from studies of the electrical properties of ice, is summarized for convenience in table 7.3.

TABLE 7.3. *Properties of orientational defects in pure ice at* -10 *°C*

Concentration of defects ($n_D = n_L$)	7×10^{15} cm^{-3}
Energy of formation of defect pair (H^f_{DL})	0·68 eV
Activation energy for diffusion (H^t_{DL})	0·235 eV
Mobility (μ_L)	2×10^{-4} cm^2 V^{-1} s^{-1}
Mobility ratio (μ_L/μ_D)	$1 \cdot 5 \pm 0 \cdot 2$

7.3. Impurities, ions and defects

Whilst most of the impurities in liquid water remain preferentially in the liquid phase when it begins to freeze, it is possible to include small amounts of some materials in true solid solution in an ice crystal. Two types of impurity, which we may call proton donors and proton acceptors, are of particular importance since they can change the balance of ion states and of orientational defects in the ice crystal. In this way they are very similar to the electron donor and electron acceptor impurities which are familiar in normal semiconductors and, as we shall see later, the analogy is quite far reaching.

The important thing about a proton donor is that it enters the ice structure substitutionally, replacing a water molecule, and possesses an extra proton. For this to occur the impurity molecule must be very nearly the same size as the water molecule and the optimum case is that of ammonia, NH_3. The incorporation of an ammonia molecule into the ice lattice is shown in a simplified way in fig. 7.4(*a*), where the real ice structure has been replaced for clarity by a two-dimensional four-bonded net. If all other molecular orientations in the crystal are held fixed, then there is clearly a D-defect generated at the impurity site. The energy of this defect is either higher or lower than that of a defect in pure ice; in the first case nearly all defects attached to impurities will be immediately liberated to diffuse away, as shown in fig. 7.4(*b*), while in the second some fraction of the defects will remain bound to the ammonia sites while some are liberated thermally. In either event, the

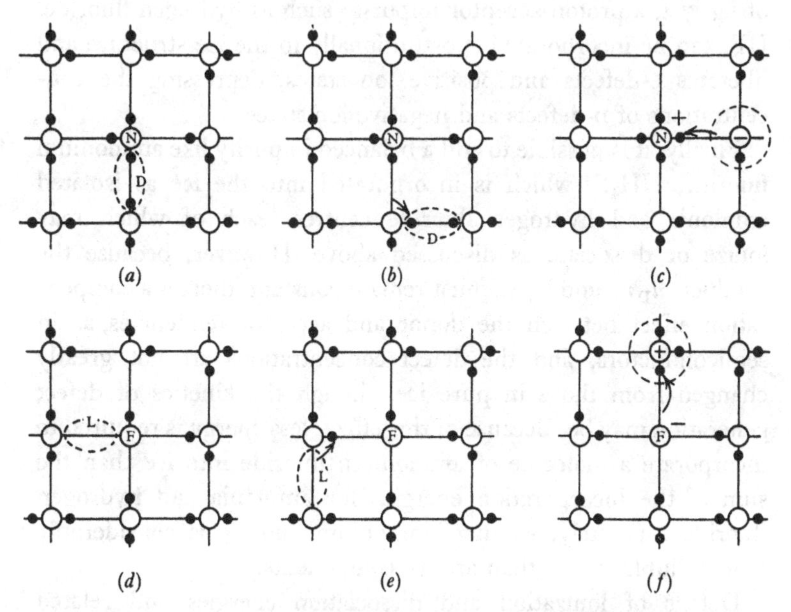

Fig. 7.4. (*a*) A proton donor impurity centre NH_3 shown schematically in a two-dimensional analogue of the ice lattice; (*b*) liberation of a D-defect by proton-jump diffusion; (*c*) subsequent liberation of a negative ion state; (*d*) a proton acceptor impurity centre HF in the same lattice; (*e*) liberation of an L-defect; (*f*) liberation of a positive ion state. Note that liberation of the ion state does not require previous dissociation of the orientational defect.

concentration n_D of D-defects in the ice crystal will be enhanced and, since the product $n_D n_L$ must remain constant, the concentration n_L of L-defects will be depressed.

Another possible modification of the donor site can take place as shown in fig. 7.4(*c*). There is always a finite possibility of a proton jump along a bond at the impurity site creating a pair of ion states. Between two water molecules the process is symmetrical but at an NH_3 impurity a jump in one sense will yield $NH_2^- + H_3O^+$ and in the other sense $NH_4^+ + OH^-$. The energies in the two cases are not the same and, in fact, the second possibility is strongly favoured. The NH_4^+ ion is fixed in the crystal, except for very slow molecular diffusion, but the OH^- state can be liberated thermally and diffuse away. Thus ammonia impurity will raise the negative ion state concentration n_- and depress the positive ion concentration n_+.

In an exactly similar way, which is illustrated in the second part of fig. 7.4, a proton acceptor impurity such as hydrogen fluoride, HF, can be incorporated substitutionally in the ice structure and liberates L-defects and positive ion states, depressing the concentrations of D-defects and negative ion states.

Finally, it is possible to add a balanced impurity like ammonium fluoride, NH_4F, which is incorporated into the ice as isolated ammonia and hydrogen fluoride centres, each of which may ionize or dissociate as discussed above. However, because the products $n_D n_L$ and $n_+ n_-$ must remain constant, there is a compensation effect between the donor and acceptor tendencies, as in semiconductors, and the defect concentrations are not greatly changed from those in pure ice, though the kinetics of defect generation may be. Because of this effect, less energy is required to incorporate a molecule of ammonium fluoride into ice than the sum of the incorporation energies for ammonia and hydrogen fluoride separately, so that ammonium fluoride is considerably more soluble in ice than are its components.

Details of ionization and dissociation energies and related quantities are found from dielectric and conductivity studies, as we shall see in chapter 9, but for convenience the experimental information is assembled in table 7.4. Statistical considerations relate the concentration n of dissociated defects to the total concentration N of impurity centres by the mass action relation

$$n^2/(N-n) = K = K_0 \exp(-H/kT), \qquad (7.5)$$

where K is the dissociation constant and H is its activation energy. The energy for liberation of an L-defect from an HF centre is very small so that K is very large and almost all these defects are freed. For HF concentrations large compared with the concentration of L-defects in pure ice, n_L is almost equal to N_{HF}. D-defects, however, are more tightly bound to NH_3 centres so that K is small and $n_D \ll N_{NH_3}$. From (7.5) then, for large ammonia concentrations, n_D varies as $(N_{NH_3})^{\frac{1}{2}}$. The ionization energies for either HF or NH_3 in ice are also considerable, though less than that of H_2O itself, so that there is relatively little ionization, in contrast to solutions in liquid water. This is due, of course, to the small effective dielectric constant of ice for fast-moving ion states, as we discussed before. Thus, for large concentrations of hydrogen fluoride or ammonia, the appropriate ion-state concentration is again proportional to the square root of impurity concentration.

TABLE 7.4. *Experimental data on behaviour of hydrogen fluoride and ammonia in ice at* $-10\ °C$

Hydrogen fluoride, HF	
Dissociation constant for H_3O^+ liberation	5×10^{10} cm^{-3} (2); 5×10^{9} cm^{-3} (3, 5)
Activation energy for H_3O^+ liberation	0·64 eV (1, 2)
Dissociation constant for L liberation	$> 10^{19}$ cm^{-3} (2)
Activation energy for L liberation	$\lesssim 10^{-2}$ eV (2)
Ammonia, NH_3	
Dissociation constant for OH^- liberation	3×10^{8} cm^{-3} (3); 2×10^{10} cm^{-3} (5)
Activation energy for OH^- liberation	0·7 eV (4)
Dissociation constant for D liberation	$\sim 10^{16}$ cm^{-3} (4)
Activation energy for D liberation	$\sim$0·3 eV (4)

(1) Steinemann (1957), (2) Jaccard (1959), (3) Iribarne *et al.* (1961), (4) Levi & Lubart (1961), (5) Levi *et al.* (1963).

7.4. Impurities and diffusion

Our discussion so far has considered the impurities to be fixed in the ice lattice but this is obviously not exactly true for they can move slowly by solid-state diffusion. Similarly individual water molecules migrate in a self-diffusion process which can be followed by using isotopically labelled molecules. The structure and mass of these are very little different from those of ordinary water molecules, so that a study of their diffusion gives information about self-diffusion in the ice crystal. The only case where this is not obviously true is for the isotopes of hydrogen, where the mass ratio to the proton is considerable, but the nature of the experimental results enables us to sidestep this difficulty.

The diffusion of the isotope ^{18}O has been studied by Kuhn & Thürkauf (1958) and by Delibaltas *et al.* (1966), ^{2}H by Kuhn &

Thürkauf (1958) and ^{3}H by Itagaki (1964), by Blicks *et al.* (1966) and by Dengel *et al.* (1963, 1966), using in all cases single crystals of ice to avoid possible grain boundary effects. The significant thing about the results of all these studies is that the diffusion coefficient of all these tracers in ice is essentially the same, about $(2 \text{ to } 3) \times 10^{-11}$ cm^2 s^{-1} at -10 °C with an activation energy of $0{\cdot}63 \pm 0{\cdot}05$ eV. This indicates quite conclusively that the diffusion mechanism is the same in each case and, by implication, the same as that for normal water molecules in the structure. Self-diffusion thus takes place by the motion of complete molecules, and the proton jumps, so significant in relaxation and conductivity processes, do not contribute.

Clarification of details of the diffusion mechanism is more difficult. Haas (1962) suggested that the most likely diffusing entity was an interstitial molecule coupled to either an L- or a D-defect, but this is disproved by the observation that the diffusion coefficient for ^{3}H is independent of the concentration of either hydrogen fluoride (Blicks *et al.* 1966) or ammonia (Dengel *et al.* 1966) over a wide range of concentrations. Onsager & Runnels (1963) considered the most likely mechanism to be that of free diffusion of interstitial water molecules with an interchange with lattice molecules only after several interstitial steps. This model is intrinsically reasonable since there are large cavities in the ice structure which can readily accommodate a water molecule, as evidenced by the structure of Ice VII, the estimated interstitial concentration at -10 °C, as we shall see later, being 10^{10}–10^{11} cm^{-3}. In addition, the model gives good agreement with measured proton spin-lattice relaxation times. A difficulty arises however because the interstitial paths for diffusion in the c-axis direction differ in cavity geometry from those for diffusion perpendicular to this direction, so that different activation energies should be expected, in contradiction with experiment. Blicks *et al.* (1966) favour a vacancy mechanism for diffusion, partly for this reason and partly because such a model can also explain the observed difference of about 10 per cent in diffusion coefficients parallel and perpendicular to the c-axis—a given number of jumps taking a molecule slightly further in the perpendicular than in the parallel direction. The formation energy for a vacancy is, as we shall see presently, about 0·53 eV, giving an estimated vacancy concentration at -10 °C of 10^{12} cm^{-3}, and the remaining 0·1 eV of the observed activation energy may be attributed to the barrier to vacancy migration, as seems very reasonable. On the basis of presently available evidence it thus seems that the vacancy mechanism for self-diffusion is most likely correct.

The only foreign molecule whose diffusion behaviour has been studied is hydrogen fluoride, and here there is some additional complication because of its low equilibrium solubility in ice (mole fraction about 4×10^{-5}). Steinemann (1957) made an estimate of 5×10^{-11} cm^2 s^{-1} for the diffusion coefficient of hydrogen fluoride in ice at -10 °C on the basis of the change with time of the properties of thin crystals containing hydrogen fluoride, though the inference was not direct. Kopp *et al.* (1965) recently made a direct measurement using the spin-lattice relaxation time for protons to measure the concentration of hydrogen fluoride in different parts of the specimen and found the very different value of $(8 \pm 5) \times 10^{-7}$ cm^2 s^{-1} at -10 °C with an activation energy of $0{\cdot}58 \pm 0{\cdot}1$ eV. Results of individual measurements in this case showed rather large scatter, as though uncontrolled imperfections such as dislocations play a considerable part in the diffusion process, but no marked anisotropy was detected.

These apparently contradictory results can perhaps be reconciled when it is realized that Steinemann's work was done with a relatively high concentration of hydrogen fluoride while, because of their much more sensitive measuring techniques, Kopp *et al.* worked with concentrations several orders of magnitude lower. Now dislocations in a crystal provide easy diffusion paths, no matter what the detailed mechanism of diffusion, because of the atomic disorder near their cores. They are, however, able to transport only very small quantities of solute because of their low cross-sectional area. Bulk diffusion by a vacancy or interstitial mechanism, on the other hand, has a large transport capacity but a much smaller diffusion coefficient. The two mechanisms are thus essentially in parallel and sensitive measurements well away from the impurity source will show a diffusion gradient characteristic of the dislocation mechanism while measurements in the high concentration region near the source detect only the normal bulk behaviour. Whilst it has not been established that this is the situation for hydrogen fluoride in ice, such a differential diffusion mechanism is known to occur in some other materials and seems to account satisfactorily for the observations.

7.5. Vacancies, interstitials and radiation damage

In any crystal there will be a certain concentration of vacancies and interstitial molecules in thermal equilibrium, the concentrations being determined by the temperature and by the energy, or more properly the enthalpy, required to produce the defect in question. In ordinary ice at atmospheric pressure the energy required to form a vacancy is essentially that necessary to remove the molecule from the bulk of the crystal and place it in a general position on the surface (at a kink in a surface step). In so doing, the number of hydrogen bonds to the molecule is reduced from four to two and the number of its neighbours of all orders is halved. As a correction we must then allow the lattice around the vacancy to relax elastically, which will lower its energy slightly. To a first approximation this vacancy energy should be equal to the sublimation energy of ice, which corresponds to removing a general surface atom to infinity, breaking in the process two hydrogen bonds and reducing the number of neighbours of all orders to zero. This energy is, as we saw in chapter 2, about 12 kcal $mole^{-1}$ or 0·5 eV. The correction for non-linear and co-operative effects in bonding, such as the increase in effective dipole moment of molecules in the crystal which we discussed in chapter 2, will raise this energy by perhaps 10 to 20 per cent, since the bonds broken in bringing the molecule to the surface are thus rather stronger than those bonding it at the surface, but elastic relaxation will reduce the energy by a roughly similar amount. In the absence of detailed calculation we may thus take the vacancy energy as 0·5 eV, which leads to an estimated vacancy concentration of 10^{12} cm^{-3} at -10 °C.

The formation of an interstitial, on the other hand, involves the transfer of a molecule from a general surface position to an interstitial position, breaking in the process two hydrogen bonds and replacing the Van der Waals interaction with a half-population of normal neighbours by the closer interactions around the interstitial site. The short-range repulsive interactions with nearest neighbours will also contribute appreciably and elastic distortion of the surrounding lattice will be important.

Because of the open, four-coordinated structure of ordinary ice, the energy involved in the creation of an interstitial molecule is not extreme. This is doubly clear from the existence of Ice VII, in which the energy penalty for the creation of a whole network of interstitials in a self-clathrate structure is more than balanced by the energy gained from the decrease in volume once the pressure exceeds 22 kbar (see chapter 3), though the phase transition I $\rightarrow$ VII would occur at a much lower pressure than this if intermediate high-pressure structures did not intervene. These energy balances have been carefully considered by Kamb (1965*b*), who finds from the experimental data that the internal energy of Ice VII at 24 kbar exceeds that of Ice I at zero pressure by about $1{\cdot}2 \pm 0{\cdot}2$ kcal mole^{-1}. This energy difference arises mainly through interactions between the two hydrogen-bonded networks of the structure, Van der Waals attractions contributing $-4{\cdot}8$ kcal mole^{-1} and repulsive overlap an estimated $+5{\cdot}4$ kcal mole^{-1}, making $+0{\cdot}6$ kcal mole^{-1}, with the remaining 0·6 kcal mole^{-1} being contributed largely by hydrogen-bond stretching of the individual frameworks. Since 1 mole of Ice VII consists of essentially $\frac{1}{2}$ mole of interstitials in $\frac{1}{2}$ mole of framework, except that the interstitials are bonded to each other, we can make a reasonable estimate of the interaction of a single interstitial with its neighbours as $+2$ to $+3$ kcal mole^{-1}. Combining this with the sublimation energy of 12 kcal mole^{-1}, we expect a total formation energy for an interstitial molecule of 14–15 kcal mole^{-1} with a resulting concentration of about 10^{10}–10^{11} cm^{-3}, which is one to two orders of magnitude less than the vacancy concentration.

Thus, although vacancies and interstitials exist in thermal equilibrium in ice, their concentrations are small even at the melting point and for this reason the density of ice crystals, measured by macroscopic means, agrees closely with the value derived from X-ray determinations of lattice constants.

To produce larger concentrations of defects we must resort to a non-equilibrium situation such as exists in a crystal bombarded by energetic radiation. If the irradiation is carried out at very low temperatures, the damage does not immediately anneal away and the different transient imperfections produced can be studied. A detailed discussion of the radio-chemistry of ice would take us too far afield and we will be content with a brief mention of some major aspects.

Irradiation of ice with γ-rays, typically from ^{60}Co, at the temperature of liquid nitrogen or liquid helium, produces hydrogen atoms and OH radicals, whose presence and behaviour can be studied by electron spin resonance methods (Siegel *et al.* 1961). In addition, if the ice contains alkali hydroxide impurities, electrons can be trapped, producing the analogue of an F-centre and giving the irradiated ice a deep blue colour (Moorthy & Weiss, 1964; Kevan, 1965). The electron trap in ice is, however, more complex than the simple cation vacancy which forms an F-centre in an alkali halide crystal. An O^- ion seems to be involved together with the stabilizing action of the alkali metal ions. Ice which has been irradiated in this way or with X-rays may show luminescent emission as it is warmed towards room temperature and from the glow curves information of the processes and defects involved can be obtained (Grossweiner & Matheson, 1954). We cannot, however, pursue these topics here.

CHAPTER 8

MECHANICAL PROPERTIES

In many parts of the world, ice is an important engineering material. Either it must serve as a substrate to bear the weight of buildings, vehicles and aircraft or else it is an obstacle which must be dug, drilled or blasted to reach something lying below. Sometimes the ice involved is merely compacted snow and sometimes it is very impure, as when formed from sea water, but often, as in glaciers, one deals with masses of almost pure ice of large grain size and with very little included air. In this latter case the properties of the whole mass can be understood quite well from fundamental studies on pure single crystals, while the former situations involve considerations peculiar to the specific environment.

The present chapter is divided into four parts. After a review of elasticity theory, we shall discuss simple elastic behaviour of single crystals of ice, then examine the relaxation processes which take place for periodically varying stresses and finally take up a brief treatment of plastic deformation, creep and related topics.

8.1. Elasticity theory

Since treatments of the elastic properties of crystals are often very brief, or at best limited to a discussion of cubic crystals, let us begin by giving a short review of this subject as it applies to hexagonal crystals like ice. We shall follow the development given by Nye (1957), to whose book the reader is referred for a fuller treatment.

The stresses acting on a solid body can be referred to rectangular axes Ox_1, Ox_2, Ox_3 and represented by the nine components σ_{ij}, where σ_{ij} is the component of force in the $+Ox_i$ direction transmitted across the face of a unit cube normal to Ox_j by the material outside the cube upon the material inside the cube. The components with $i = j$ are normal stresses and those with $i \neq j$ are shear stresses. The nine components σ_{ij} form a second rank tensor, which is symmetric:

$$\sigma_{ij} = \sigma_{ji}. \tag{8.1}$$

The components of strain can be related to the displacement u_i suffered by material at x_j by the relation

$$e_{ij} = \frac{\partial u_i}{\partial x_j} \tag{8.2}$$

and the nine components e_{ij} form a second rank tensor. However, a simple rotation of the whole crystal without any strain gives a set of non-zero e_{ij} which are antisymmetric, so that the true strain is represented by the symmetric part of e_{ij}, given by

$$\epsilon_{ij} = \tfrac{1}{2}(e_{ij}+e_{ji}) \tag{8.3}$$

These ϵ_{ij} form a symmetric second rank tensor called the strain tensor.

Ordinary elasticity theory is based upon Hooke's law: the assumption that there is a linear relationship between stress and strain. For a crystal this takes the form

$$\epsilon_{ij} = s_{ijkl}\,\sigma_{kl} \tag{8.4}$$

where the eighty-one components of the tensor s_{ijkl} are the compliances of the crystal and, for all of this section, the summation convention over repeated indices is assumed. The inverse relation has the form

$$\sigma_{ij} = c_{ijkl}\epsilon_{kl}, \tag{8.5}$$

where the eighty-one tensor components c_{ijkl} are called the stiffness constants of the crystal.

These equations can be considerably simplified by symmetry considerations, which require

$$s_{ijkl} = s_{jikl}, \quad s_{ijkl} = s_{ijlk}, \quad s_{ijkl} = s_{klij} \tag{8.6}$$

together with similar relations for c_{ijkl}. These, together with the symmetry of σ_{ij} or ϵ_{ij}, reduce the number of independent components of σ_{ij} or ϵ_{ij} to six and the number of independent s_{ijkl} or c_{ijkl} to twenty-one. A compact matrix rotation can then be introduced by abbreviating the first two subscripts of s_{ijkl} to a single subscript according to the scheme:

Tensor notation	11	22	33	23, 32	31, 13	12, 21	(8.7)
Matrix notation	1	2	3	4	5	6	

The same rule is used to convert the last pair of subscripts and, at the same time, we introduce a factor 2 or 4 according to

$$\left.\begin{aligned} s_{ijkl} &= s_{mn} \quad \text{when both } m \text{ and } n \text{ are 1, 2 or 3,} \\ 2s_{ijkl} &= s_{mn} \quad \text{when either } m \text{ or } n \text{ is 4, 5 or 6,} \\ 4s_{ijkl} &= s_{mn} \quad \text{when both } m \text{ and } n \text{ are 4, 5 or 6.} \end{aligned}\right\} \tag{8.8}$$

The strain components ϵ_{ij} are converted to single-index quantities ϵ_m using (8.7) and, instead of (8.8), the rules

$$\left.\begin{aligned} \epsilon_{ij} &= \epsilon_m \quad \text{if} \quad m = 1, 2 \text{ or } 3, \\ 2\epsilon_{ij} &= \epsilon_m \quad \text{if} \quad m = 4, 5 \text{ or } 6. \end{aligned}\right\} \tag{8.9}$$

The compliances c_{ijkl} and stress components σ_{ij} are converted to c_{mn} and σ_m by the same rule (8.7) but no numerical factors as in (8.8) or (8.9) are required.

The slight complication of these conversions results in greater simplicity for the final results

$$\epsilon_i = s_{ij}\sigma_j, \quad \sigma_i = c_{ij}\epsilon_j \quad (i = 1, 2, \ldots, 6), \tag{8.10}$$

and we still have the symmetry relations

$$c_{ij} = c_{ji}, \quad s_{ij} = s_{ji}, \tag{8.11}$$

though the c_{ij} and s_{ij} no longer transform as tensors but are simply matrix coefficients. In terms of these new variables, the strain energy density is simply

$$U = \tfrac{1}{2}c_{ij}\epsilon_i\epsilon_j. \tag{8.12}$$

All these results apply to a completely general triclinic crystal system whose elastic properties are expressed by the twenty-one independent quantities c_{ij} or s_{ij}. For crystals of higher symmetry there are further relations between the c_{ij} or s_{ij} which reduce their number still further. For the hexagonal and cubic systems these relations are illustrated in fig. 8.1, together with similar relations for a completely isotropic, non-crystalline material. It can be seen that for a hexagonal crystal like ice there are only five non-zero independent elastic constants s_{11}, s_{12}, s_{13}, s_{33} and s_{44}, or the corresponding c_{ij}.

The c_{ij} and s_{ij} are related quite generally by

$$c_{ij}s_{jk} = \delta_{ik}. \tag{8.13}$$

For the hexagonal system this leads to the explicit results

$$c_{11}+c_{12} = s_{33}/s, \quad c_{11}-c_{12} = 1/(s_{11}-s_{12}), \quad c_{13} = -s_{13}/s,$$
$$c_{33} = (s_{11}+s_{12})/s, \quad c_{44} = 1/s_{44}, \tag{8.14}$$
$$s = s_{33}(s_{11}+s_{12})-2s_{13}^2,$$

and the same relations hold if all c_{ij} and s_{ij} are interchanged.

Since, as we shall see presently, ice is not very anisotropic in its elastic behaviour (in sharp distinction with its plastic flow), it is sometimes useful to approximate its behaviour by the common

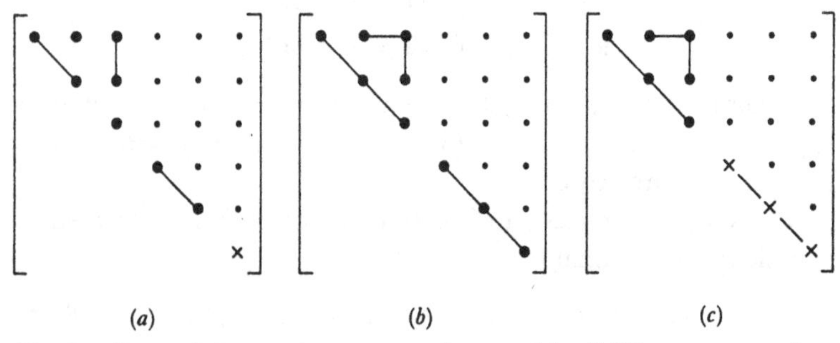

Fig. 8.1. Form of the matrices s_{ij} or c_{ij} for materials of different symmetries: (*a*) hexagonal, (*b*) cubic, (*c*) isotropic. All matrices are symmetric about the leading diagonal and the following symbols are used: ·, zero component; •, non-zero component; •—•, equal components; ×, $2(s_{11}-s_{12})$ for s_{ij} or $\frac{1}{2}(c_{11}-c_{12})$ for c_{ij} (after Nye, 1957).

elastic moduli used for isotropic materials: the Young's modulus E, rigidity modulus G and Poisson's ratio γ. The appropriate relations for a truly isotropic material are

$$E = 1/s_{11}, \quad \gamma = s_{12}/s_{11}, \quad G = 1/[2(s_{11}-s_{12})], \tag{8.15}$$

from which, of course,

$$G = E/[2(1+\gamma)]. \tag{8.16}$$

Alternatively we may introduce the elastic constants λ and μ, used in some treatments of elasticity, through the relations

$$c_{11} = 2\mu+\lambda, \quad c_{12} = \lambda. \tag{8.17}$$

These equations are not exactly those for a finely polycrystalline mass of ice but approximate to them. The more complex expressions obtained when a correct average is made over all orientations in such an aggregate are given by Penny (1948).

8.2. Elastic properties of ice

Since the description of the elastic properties of ice requires five independent parameters, it is necessary to measure five physical quantities in order to determine them all. In principle we could make a direct static measurement by applying stresses to the crystal, measuring the strains and using (8.10). This is not a very good method in practice, however, because the ice creeps slightly even under very small loads. It has been used by Gold (1958) and others, as we shall discuss presently, to measure bulk properties like Young's modulus and Poisson's ratio, but most measurements on single crystals have been dynamic.

In dynamic methods the quantity measured is essentially the velocity of propagation for a particular type of elastic wave in a particular crystal direction and, from a suitably chosen set of such measurements, the s_{ij} or c_{ij} can be deduced. Generally one of these sets of coefficients is derived directly from the measurements, while the other set, found from (8.14), has a rather lower accuracy. Such dynamic measurements have been made by Jona & Scherrer (1952), who excited the crystal elastically with ultrasonic waves of frequency 15–18 MHz and examined the nodal pattern of these waves by the diffraction effects produced upon monochromatic light passing through the crystal. The linear relation between frequency and wavelength determined in this way for propagation in various directions gave all five c_{ij} at a temperature of -16 °C. Zarembovitch & Kahane (1964) later used a similar method at frequencies 6–14 MHz to examine the behaviour of c_{11} and c_{33} over the temperature range -2 to -180 °C, their results agreeing well with those of Jona and Scherrer in the region of overlap. A related method, but using much lower frequencies (10–50 kHz) and determining the standing wave pattern by direct observation in polarized light, was used by Bass *et al.* (1957) and analysed in such a way as to give the s_{ij} rather than the c_{ij}. The temperature range covered was -2 to -30 °C. Green & Mackinnon (1956) and Bogorodskii (1964), on the other hand, used a direct pulse propagation method to find the c_{ij}. The results of these measurements at -16 °C are compared in table 8.1, from which it can be seen that there is very substantial agreement between them all, after allowing for the different orders of accuracy arising through conversion between c_{ij} and s_{ij}. The variation of the elastic parameters with temperature is appreciable, though not extreme, and is shown in figs. 8.2 and 8.3. Bogorodskii (1964) found a much greater variation in the range 0 to −20 °C but this might be viewed with reserve because of the lower accuracy of his measurements.

TABLE 8.1. *The elastic parameters of ice at* −16 °C

	Jona & Scherrer (1952)	Green & MacKinnon (1956)	Bass *et al.* (1957)	Zarembovitch & Kahane (1964)	Penny (1948)
	Elastic compliances s_{ij} in units of 10^{-12} cm² dyne⁻¹				
s_{11}	10·4±0·3	—	10·13±0·05	—	9·9
s_{12}	−4·3±0·3	—	−4·16±0·15	—	−4·0
s_{13}	−2·4±0·1	—	−1·93±0·21	—	−2·5
s_{33}	8·5±0·4	—	8·28±0·04	—	8·4
s_{44}	31·4±0·3	—	32·65±0·15	—	31·2
	Elastic stiffnesses c_{ij} in units of 10^{10} dyne cm⁻²				
c_{11}	13·85±0·08	13·33±1·98	13·3±0·8	13·94±0·09	15·2
c_{12}	7·07±0·12	6·03±0·72	6·3±0·8	—	8·0
c_{13}	5·81±0·16	5·08±0·72	4·6±0·9	—	7·0
c_{33}	14·99±0·08	14·28±0·08	14·2±0·7	14·98±0·01	16·2
c_{44}	3·19±0·03	3·26±0·08	3·06±0·015	—	3·2

In table 8.1 we have also shown the results of a remarkable calculation made by Penny (1948) at a time when the only experimental data available were for the Young's modulus and Poisson's ratio of polycrystalline ice samples. By making the assumption that the symmetry of the elastic properties would not be greatly affected if the statistically distributed proton positions were replaced by single protons at the mid-point of each bond (the Barnes model for ice), and by neglecting the small deviations from exact tetrahedral symmetry of oxygen positions, she was able to deduce four relations between the six elastic parameters of ice. This reduced the total number of independent parameters to two, which could be evaluated from the available experimental results. As can be seen from table 8.1 the agreement between her calculated values and those found by later direct experiments is remarkably good.

It is interesting to make an estimate of the extent to which the elastic properties of ice are anisotropic. This can be done with the help of fig. 8.1 and the figures from table 8.1. For an isotropic material $c_{11} = c_{33}$ while for ice these quantities differ by about 10 per cent. Similarly we should have $c_{44} = \frac{1}{2}(c_{11} - c_{12})$ while for ice there is a deviation of about 10 per cent. Finally for an isotropic

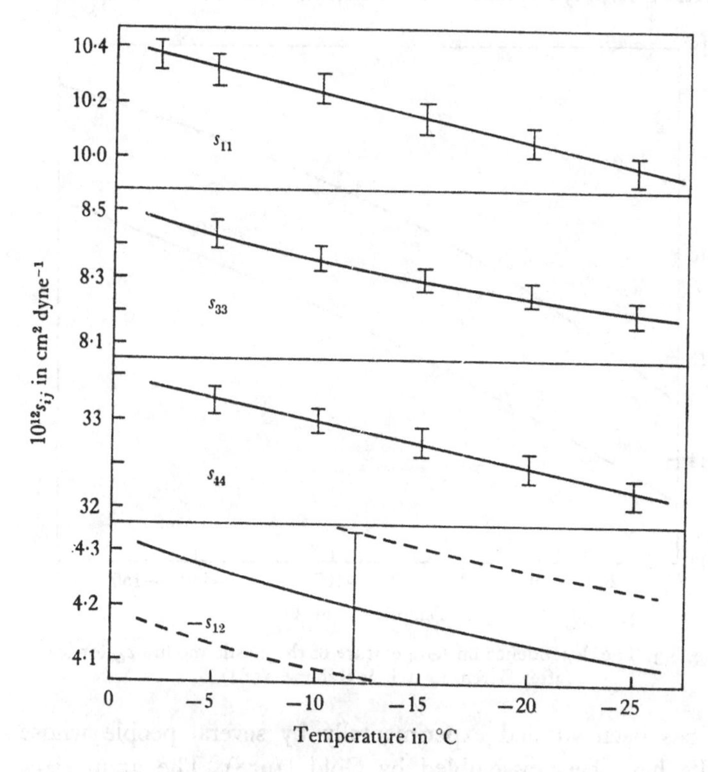

Fig. 8.2. The dependence on temperature of the elastic coefficients s_{ij} for ice (after Bass *et al.* 1957).

material $c_{12} = c_{13}$ and this relation is unsatisfied by about 30 per cent. We should therefore expect the elastic anisotropy of ice to range from about 10 to 30 per cent, depending upon the stress system involved. The coefficient s_{11} is of particular interest since, from (8.15), it is simply the reciprocal of the Young's modulus in the case of an isotropic material. For ice, when the effective s'_{11} is evaluated about axes which have been rotated through an angle θ

away from the c-axis direction, the measured value is (Bass *et al.* 1957)

$$s'_{11} = s_{33}\cos^4\theta + s_{11}\sin^4\theta + (s_{44} + 2s_{13})\cos^2\theta\sin^2\theta. \quad (8.18)$$

The effective Young's modulus $E' = 1/s'_{11}$ is plotted as a function of orientation in fig. 8.4.

Since it is often of practical consequence to know the elastic behaviour of polycrystalline ice, particularly under static loading,

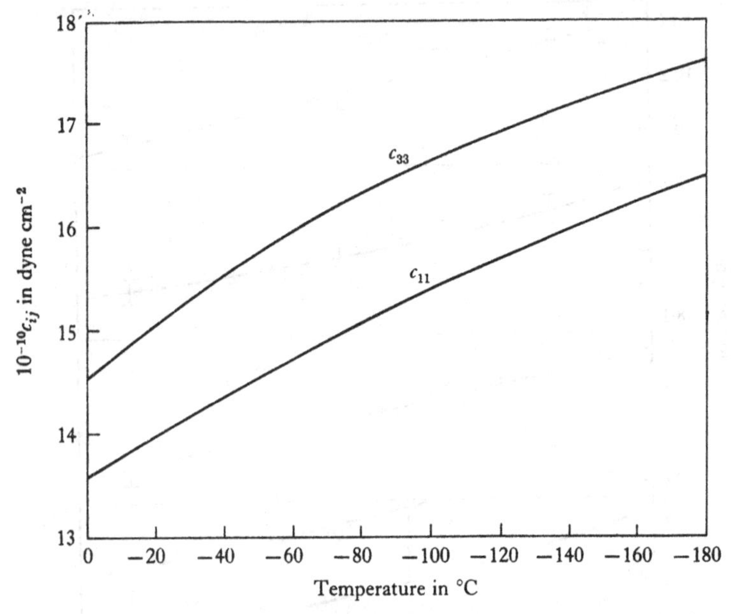

Fig. 8.3. The dependence on temperature of the elastic moduli c_{ij} for ice (after Zarembovitch & Kahane, 1964).

this has been studied experimentally by several people whose results have been assembled by Gold (1958). The grain sizes studied ranged from a few millimetres to a few centimetres and, because of the mechanism of freezing, there was generally a preferred crystal orientation and, understandably, considerable scatter of results. Gold found that, if the stress was less than 10 kg cm^{-2} and was applied for less than about 10 s, the ice behaved elastically. The results are summarized in table 8.2 and agree well with those calculated from table 8.1, assuming a random distribution of grains. There are, however, certain anomalies in the elastic behaviour of polycrystalline ice above about −40 °C which are apparently associated with grain boundary slip. From direct measurements of the eigenfrequencies of vibrating bars (Kuroiwa & Yamaji, 1959) or from the behaviour shown in fig. 8.3, the three moduli E, K and G increase by about 20 per cent as the temperature falls from 0 to − 180 °C.

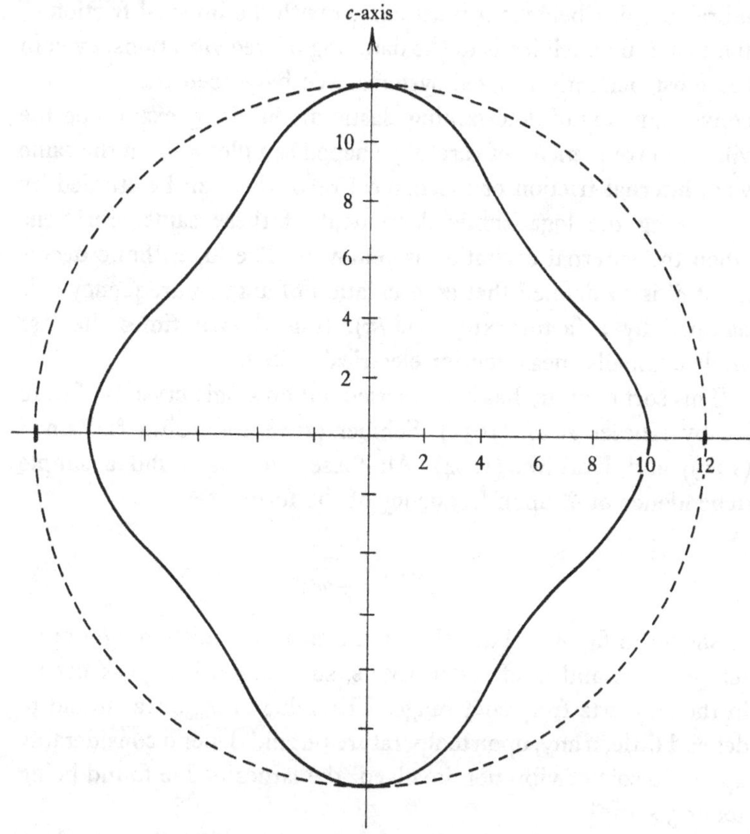

Fig. 8.4. The effective Young's modulus $E' = 1/s'_{11}$ as a function of stress orientation. Units are 10^{10} dyne cm^{-2}.

TABLE 8.2. *Elastic properties of polycrystalline ice at* − 5 °C (*Gold*, 1958)

Young's modulus (E)	$(8{\cdot}9–9{\cdot}9) \times 10^{10}$ dyne cm^{-2}
Rigidity modulus (G)	$(3{\cdot}4–3{\cdot}8) \times 10^{10}$ dyne cm^{-2}
Bulk modulus (K)	$(8{\cdot}3–11{\cdot}3) \times 10^{10}$ dyne cm^{-2}
Poisson's ratio (γ)	0·31–0·36

8.3. Mechanical relaxation

Most real systems, as distinct from idealized assemblies of point masses linked by two-body forces, show deviations from the simple behaviour discussed in the previous sections. One of these types of more complex behaviour is associated with the internal friction of the material which leads to the damping of free vibrations, even in the most perfectly isolated systems. We have seen that the most convenient way of determining elastic moduli is by examining the vibration frequencies of carefully shaped samples and, in the same way, internal friction or mechanical relaxation can be studied by examining the logarithmic decrement of these same vibrations when the external excitation is removed. The logarithmic decrement δ' is so defined that an oscillation of angular frequency ω is damped by a factor $\exp(-\omega t\delta'/\pi)$. It is thus π times the loss angle δ usually measured for electrical systems.

This sort of study has been carried out on single crystals of pure ice by Kneser *et al.* (1955), Schiller (1958), Kuroiwa & Yamaji (1959) and Kuroiwa (1964). All these workers found a simple dependence of δ' upon frequency of the form

$$\delta' = \delta'_{\text{max}} \frac{2\omega\tau}{1+\omega^2\tau^2} \tag{8.19}$$

as shown in fig. 8.5. The relaxation time τ depends strongly upon temperature and is of order 10^{-4} s, so that the loss peak occurs in the kilohertz frequency range. The value of δ'_{max} was found to depend little, if any, upon temperature but did depend considerably upon the sort of vibration involved, the largest value found being about 3×10^{-2}.

Before going on to consider these results in detail, let us look briefly at the way in which this sort of mechanical relaxation arises. The fact that there is any mechanical damping at all indicates that some process is taking place which is out of phase with the macroscopic elastic strains. This sort of thing happens, for instance, in ordinary viscous damping in which the forces in the viscous medium are proportional to velocity gradient rather than to displacement. The present phenomenon is not, however, a viscous damping process because the magnitude of the damping is greatest at a particular frequency $\omega = 1/\tau$. This behaviour is characteristic of a situation in which there are two possible configurations of the material and a characteristic transition time τ between them. If $\omega \ll 1/\tau$ then the configuration change is almost in phase with the stress causing it and the loss is very small. If $\omega \gg 1/\tau$ then there is not enough time during one period for any appreciable configuration change at all. Losses are thus confined to a range of frequencies where $\omega \sim 1/\tau$.

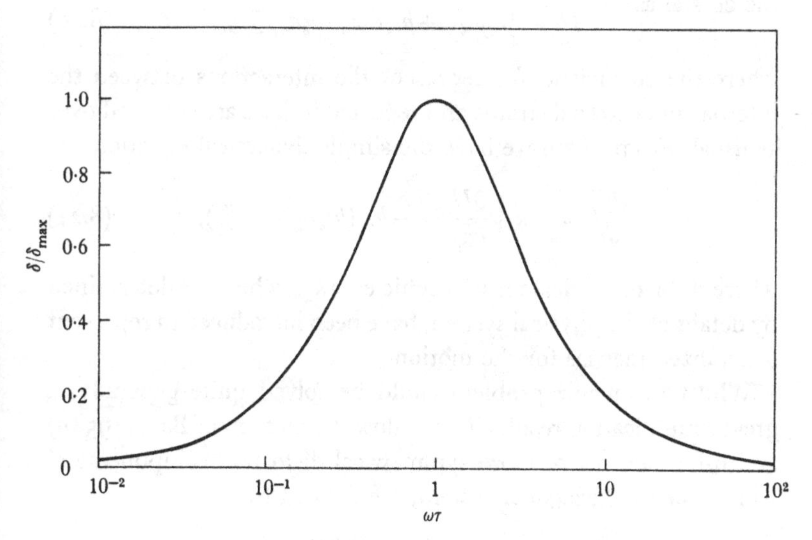

Fig. 8.5. The form of the Debye loss curve.

Whilst there are, as we shall see later, relaxation phenomena associated with grain boundaries and other gross crystal imperfections, the damping with which we are at present concerned seems to be an intrinsic property of single-crystal ice. The magnitude of the decrement δ'_{max} is little affected by the history or purity of the ice, which suggests that relaxation may be associated with a basic parameter, like the orientational ordering of water molecules, rather than with any sort of defect. This is, in fact, a correct conclusion, as we shall see later. The characteristic time τ does, however, depend markedly upon impurity content, which implies that the mechanism by which water molecules may change their orientation involves defects introduced by these impurities.

To put these ideas in formal terms let us introduce, as well as the six components ϵ_i of the elastic strain tensor, a set of strain components ξ_α for the internal co-ordinates. The Roman index i runs from 1 to 6, while the Greek index α runs from 1 to some integer large enough to specify all the internal variables of the crystal, but in fact only those variables specifying the orientations of the water molecules will ultimately be involved. As a simple generalization of (8.12) we can now write the total strain energy of the crystal as

$$U = \tfrac{1}{2}c_{ij}\epsilon_i\epsilon_j + b_{i\alpha}\epsilon_i\xi_\alpha + \tfrac{1}{2}a_{\alpha\beta}\xi_\alpha\xi_\beta, \tag{8.20}$$

where the coefficients $b_{i\alpha}$ represent the interactions between the internal and external strains and repeated indices are summed over as usual. From (8.20) we have the simple dynamical equation

$$\frac{d\xi_\alpha}{dt} = -k_{\alpha\beta}\frac{\partial U}{\partial \xi_\beta} = -k_{\alpha\beta}(b_{i\beta}\epsilon_i + a_{\gamma\beta}\xi_\gamma), \tag{8.21}$$

where the symmetrical tensor coefficients $k_{\alpha\beta}$, which are determined by details of the physical system, have been introduced to represent generalized inertias for the motion.

Whilst the whole problem could be solved quite generally, a great simplification results if we follow the method of Bass (1958*a*) and use a new set of more symmetrical distortion components ϵ_i' which, for the hexagonal system, are defined by

$$\left.\begin{aligned} \epsilon_1' &= (\epsilon_1+\epsilon_2+\epsilon_3), \quad \epsilon_2' = (1/\sqrt{3})(\epsilon_1+\epsilon_2-2\epsilon_3), \\ \epsilon_3' &= (\epsilon_1-\epsilon_2), \quad \epsilon_4' = \epsilon_4, \quad \epsilon_5' = \epsilon_5, \quad \epsilon_6' = \epsilon_6. \end{aligned}\right\} \tag{8.22}$$

Then ϵ_1' is a simple dilatation of the crystal and the remaining ϵ_i' are distortions at constant volume. ϵ_2' is a distortion with rotational symmetry about the c-axis and ϵ_3' to ϵ_6' represent various plane shears. For these strains and the similarly defined stresses we have new stiffness coefficients which can be shown to be

$$\left.\begin{aligned} c_{11}' &= \tfrac{1}{9}(2c_{11}+2c_{12}+4c_{13}+c_{33}), \\ c_{12}' &= (1/3\sqrt{3})(c_{11}+c_{12}-c_{13}-c_{23}), \\ c_{22}' &= \tfrac{1}{6}(c_{11}+c_{12}-4c_{13}+2c_{33}), \\ c_{33}' &= c_{66}' = \tfrac{1}{2}(c_{11}-c_{12}), \\ c_{44}' &= c_{55}' = c_{44}. \end{aligned}\right\} \tag{8.23}$$

When we come to consider the internal strains ξ_α we make the assumption that the only co-ordinates of importance are those specifying proton positions or, equivalently, the orientations of water molecules. There are two non-equivalent molecular positions in the ice structure and for each of these the molecular dipole can be oriented in one of two senses along three possible directions. In a nearest-neighbour calculation, once the dipole axes are specified, the sense of the dipole on the central molecule determines that of its neighbours and does not affect the energy. We are therefore left with only six internal variables which specify, for each of the two molecular positions, the fraction of molecules having their dipole directed along each of the three possible dipole axis directions. Since the sum of these probabilities must be unity for each type of molecular site, there are only four independent co-ordinates and, if the six internal strains ξ'_i are defined in the same way as (8.22), then both ξ'_1 and ξ'_2 are found to vanish.

With this definition of the internal co-ordinates, symmetry then requires that

$$\left.\begin{aligned} r_{ij} &= 0 \quad (i \neq j \quad \text{and} \quad i,j = 3, 4, 5, 6),\\ r_{33} &= r_{66}, \quad r_{44} = r_{55}, \end{aligned}\right\} \tag{8.24}$$

where r represents a', b', c' or k'. With these new variables and coefficients, equations (8.20) and (8.21) remain identical in form.

For the case in which we are interested, ϵ'_i has a time variation $\exp(i\omega t)$ which leads to a similar variation for the ξ'_i. These internal variables, with the help of (8.24), are then easily eliminated from (8.20) and (8.21), leading to a set of uncoupled equations which can be solved separately to give, for $i = 3, 4, 5, 6$,

$$\sigma'_i = \frac{\partial U}{\partial \epsilon'_i} = \left(c'_{ii} - \frac{b'^2_{ii}}{a'_{ii}} \frac{1}{1+i\omega\tau_i}\right)\epsilon'_i, \tag{8.25}$$

with
$$\tau_i = (k'_{ii}a'_{ii})^{-1}. \tag{8.26}$$

In both these equations there is no summation over the repeated index i.

This treatment thus leads to a complex effective stiffness constant which introduces a phase lag between σ'_i and ϵ'_i and is thus responsible for the observed damping. The logarithmic decrement for oscillation mode i is defined by

$$\delta' = \pi \frac{\mathrm{Im}(\sigma_i'/\epsilon_i')}{\mathrm{Re}(\sigma_i'/\epsilon_i')} = \pi \frac{b_{ii}'^2}{a_{ii}'} \frac{\omega\tau_i}{c_{ii}'(1+\omega^2\tau_i^2) - b_{ii}'^2/a_{ii}'}, \tag{8.27}$$

which, if δ_i' is small so that $b_{ii}'^2/a_{ii}' \ll 1$, has the observed simple form (8.19) with the maximum of δ_i' at a frequency for which $\omega\tau_i = 1$.

Another interesting form of solution can be obtained from the result (8.25). This equation, we recall, gives the stress component σ_i' required to excite a strain ϵ_i' in the mode i at a frequency ω. Suppose instead that we cause a step-function strain ϵ_i' for this particular deformation mode at time $t = 0$ and ask what is the stress as a function of time. Dropping the primes and subscripts for simplicity, let ϵ_0 be the magnitude of the strain applied at $t = 0$, then its Fourier components are simply

$$\epsilon(\omega) = (2\pi)^{-\frac{1}{2}} \int_0^\infty \exp(-i\omega t)\, \epsilon_0\, dt = (2\pi)^{-\frac{1}{2}} (i\omega)^{-1} \epsilon_0, \tag{8.28}$$

and so, from (8.25),

$$\sigma(t) = (2\pi)^{-\frac{1}{2}} \int_{-\infty}^{\infty} \left[c - \frac{b^2}{a(1+i\omega\tau)}\right] \epsilon(\omega) \exp(i\omega t)\, d\omega. \tag{8.29}$$

This is simply evaluated as a contour integral and gives the result

$$\sigma(t) = \left[c - \frac{b^2}{a}\{1 - \exp(-t/\tau)\}\right] \epsilon_0, \tag{8.30}$$

so that the stress relaxes from the value $c\epsilon_0$ towards the value $(c - a^{-1}b^2)\,\epsilon_0$ with a characteristic relaxation time τ. (Note that the convergence of the integrals in (8.28) and (8.29) can be effected by allowing ω to have an infinitesimal imaginary part of appropriate sign.)

The theory we have outlined is based upon the assumption that it is changes in the orientation of the water molecules in the crystal which cause the absorption of energy. These orientations are normally random, as we saw in chapter 2, but, under the influence of strain, some dipole orientations may become energetically favoured, leading to a coupling between the external strains ϵ_i' and the internal strains ξ_i'. The coupling coefficients have been evaluated by

Bass (1958*a*), using a simplified model for the lattice in which each O–H...O bond is replaced by a point dipole at the midpoint of the bond. On this approximation the total lattice energy is independent of the orientational order, as is assumed in the Pauling model. The internal constant a'_{ii} is found to depend upon temperature, since an entropy term is involved, and the final value of δ'_{max} is approximately inversely proportional to absolute temperature.

One of the first results of this treatment is, as we have seen, that the internal co-ordinates ξ'_1 and ξ'_2 vanish exactly, implying that there is no mechanical damping from this cause associated with oscillations which are simple dilatations (if they could be produced) or with those whose strain field has axial symmetry about the *c*-axis. This is verified experimentally, in that the damping of axially symmetric modes is an order of magnitude less than that of other modes. The measured decrements and predicted ratios are shown in table 8.3. The absolute calculated values depend to some extent upon the numerical values assumed for the calculation but, for reasonable assumptions, δ'_{max} ranges from 4×10^{-2} to 3×10^{-1}, in rough agreement with experiment. The measurements also show a shift of about 8 per cent in the resonance frequency when τ^{-1} is moved through it by a variation of the temperature, again in agreement with theoretical predictions.

TABLE 8.3. *Logarithmic decrement δ'_{max} for different vibrational modes in ice (Schiller, 1958 B;ass, 1958a, b)*

Orientation	Wave type	δ'_{max} (experimental)	Theoretical relative value
$\parallel_c$	Transverse	3.0×10^{-2}	1
	Longitudinal	0.35×10^{-2}	0
$\perp_c$	Transverse	3.3×10^{-2}	1
	Longitudinal	2.5×10^{-2}	0.7

The interpretation of measured relaxation phenomena in terms of rotation of water molecules in the ice structure thus seems to account well for the observations. The only other internal coordinates available in pure ice crystals are those associated with crystal defects such as dislocations or broken bonds. If broken bonds are assumed to be present in thermal equilibrium, then their contribution to damping can be calculated (Bass, 1958*b*) and is of order 10^{-8}, which is far less than measured values. In addition their contribution δ'_{max} should have a strong temperature dependence and be equally present for axially symmetric modes, in contradiction with experiment. Similar objections apply to an interpretation in terms of dislocations.

Although lattice imperfections are thus not directly responsible for mechanical relaxation phenomena, they do play an important role by providing a mechanism for changing the average orientation of water molecules in the crystal. A single molecule in the crystal cannot rotate by itself without violating the normal bonding requirements with neighbouring molecules and, in fact, the smallest group of molecules which can make a collective rotation without breaking bonds is the set of six forming a puckered hexagonal ring. A simplified version of this geometry is shown in fig. 8.6(*a*). For the particular set of molecular orientations shown, which might be described as a clockwise ring, we can reach another configuration by rotating each molecule through an angle $2\pi/3$ about its bond perpendicular to the ring, so arriving at the anticlockwise ring shown in fig. 8.6(*b*), all bonds to neighbouring molecules remaining unchanged. This sort of ring behaviour can occur with larger groups of molecules as well, and indeed must involve a larger group when the proton configuration is less simple than that discussed above. Such ring interchanges do not, however, lead to any change in the relevant internal co-ordinate; each bond in fig. 8.6(*a*) is exactly balanced by an oppositely directed bond and the same remains true in fig. 8.6(*b*). The same conclusion follows when the co-operative rotation of larger groups of molecules is considered.

A real ice crystal at a finite temperature is, however, not perfect but contains an equilibrium concentration of point defects, as discussed in chapter 7. The most important of these in pure ice, from our present point of view, are the D- and L-orientational defects, since the product of their concentration and mobility is about 100 times greater than the same product for ion states, so that they provide the dominant relaxation mechanism.

The way in which a D-defect allows changes in molecular

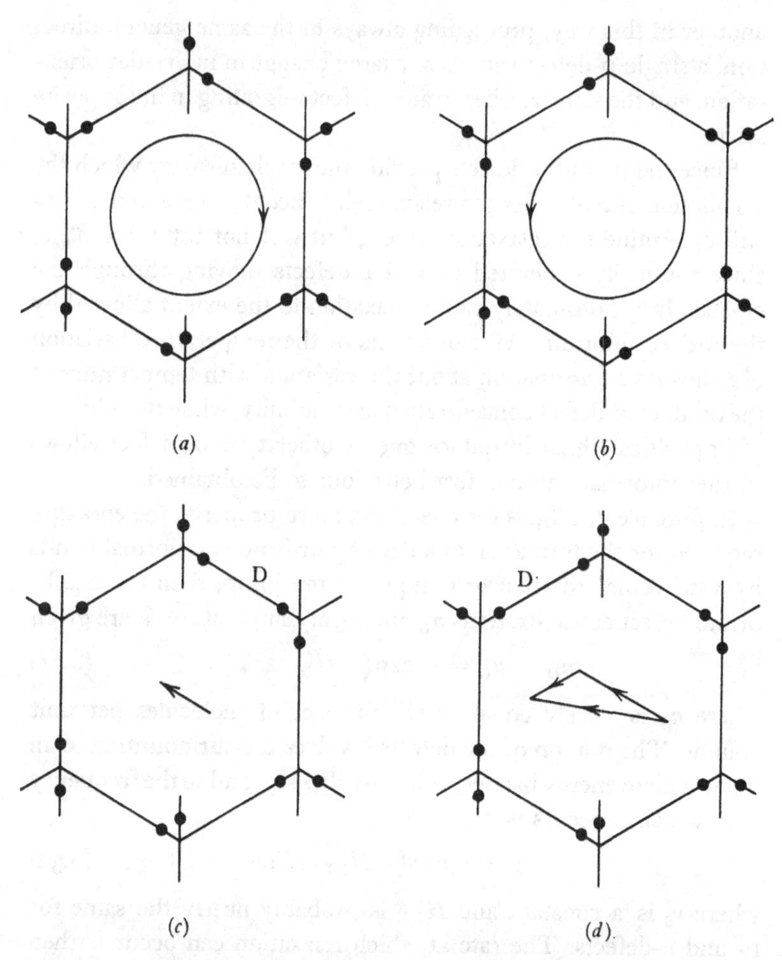

Fig. 8.6. Possible proton configurations and defect structures in ice. (*a*) A hexagonal ring with clockwise bonds, (*b*) a hexagonal ring with anticlockwise bonds, (*c*) net bond orientation in a ring containing a D-defect, (*d*) resultant bond orientation after motion of the defect.

orientation is shown in fig. 8.6(*c*), where we have a clockwise ring containing a single D-defect. If we sum the vectors associated with each bond in the ring then we are left with the single vector shown, corresponding to the unbalanced bond opposite to the D-defect. A simple jump of the defect to a neighbouring bond rotates this resultant and increases it, as shown in fig. 8.6(*d*), leaving all other bonds in the crystal unchanged. By moving from one ring to another in this way, proceeding always in the same general direction, a single D-defect can cause a large change in molecular orientation, and the same applies to an L-defect migrating in the opposite sense.

Since the D- and L-defects provide the mechanism by which the orientation of molecules can relax, their concentration and mobility will determine the relaxation time τ, but will not influence δ'_{max}, since thermally generated D- and L-defects moving through the crystal always ultimately cause relaxation to the extent allowed by thermal equilibrium. Measurements of the temperature variation of τ thus give information about the variation with temperature of the product of defect concentration and mobility, while the addition of impurities which introduce one or other type of defect allows further information on defect behaviour to be obtained.

In pure ice, if H^f_{DL} is the energy or, more properly, the enthalpy required for the formation of a defect pair from two normal bonds by a molecular rotation or oblique proton jump, then the equilibrium defect concentrations n_D and n_L at temperature T are given by

$$n_D = n_L = n_S \exp(-H^f_{DL}/2kT), \qquad (8.31)$$

where n_S is closely equal to the number of molecules per unit volume. The motion of the defects involves the surmounting of an intermediate energy barrier, of height H^t_{DL} say, and so the frequency with which it occurs is

$$\nu = \nu_0 \exp(-H^t_{DL}/kT), \qquad (8.32)$$

where ν_0 is a constant and H^t_{DL} is probably nearly the same for D- and L-defects. The rate at which relaxation can occur is then τ^{-1}, where

$$\tau = A[(n_D+n_L)\nu]^{-1} = \tau_0 \exp[(H^t_{DL}+\tfrac{1}{2}H^f_{DL})/kT] = \tau_0 \exp(H/kT). \qquad (8.33)$$

Experiments on nominally pure single crystals of both H_2O and D_2O ice verify the general form of (8.33) but there is considerable disagreement as to the numerical quantities involved, the total measured activation energy ranging from 0·26–0·58 eV according to different workers. These differences are, however, readily explained in terms of the presence of minute amounts of impurity, as was indeed recognized by Kneser *et al.* (1955) in their original work. Concentrations of as little as a few parts per million of halides like sodium chloride, hydrogen chloride, hydrogen fluoride or ammonium fluoride are found to reduce the activation energy markedly (Walz & Magun, 1959; Schulz, 1961; Kuroiwa, 1964). The most careful recent values are probably those of Kuroiwa (1964), who finds for pure H_2O ice

$$\tau_0 = 6{\cdot}9 \times 10^{-16}\ \mathrm{s}, \quad H = 0{\cdot}57\ \mathrm{eV}, \tag{8.34}$$

and for pure D_2O ice

$$\tau_0 = 1{\cdot}04 \times 10^{-5}\ \mathrm{s}, \quad H = 0{\cdot}575\ \mathrm{eV}. \tag{8.35}$$

This shift of the relaxation peak for D_2O towards lower frequencies, by about a factor 2, is supported by the previous work of Woerner & Magun (1959) and the activation energies agree with the best of the earlier determinations.

The effects of impurities like ammonia or hydrogen fluoride on the relaxation behaviour can be readily understood in terms of our discussion in chapter 7. Each of these materials introduces D- or L-defects and thus, by (8.33), reduces the relaxation time τ. At the same time the component $H^{\mathrm{f}}_{\mathrm{DL}}$ of the activation energy is reduced towards the value $H^{\mathrm{f}}_{\mathrm{D}}$ or $H^{\mathrm{f}}_{\mathrm{L}}$ appropriate for liberation of the corresponding defect from its parent impurity. This reduction is considerable and, for hydrogen fluoride in relatively large concentration, the total activation energy H approaches $H^{\mathrm{t}}_{\mathrm{DL}}$, which is only 0·235 eV. We shall not discuss these impurity effects in detail here, since essentially the same phenomena occur in the dielectric relaxation of ice, a topic which we take up in chapter 9 and which has been the subject of a much wider range of experimental studies.

Before leaving the subject of mechanical relaxation it is interesting to look at the behaviour of polycrystalline ice specimens. In single crystals, as we have seen, the contribution of dislocations and other imperfections is negligible, so that the relaxation is characterized by a single relaxation time associated with molecular reorientation. For polycrystalline specimens, however, the grain boundaries introduce their own characteristic relaxation processes which are sufficiently strong to be distinguished above the intrinsic molecular relaxation.

Figure 8.7 shows the relaxation behaviour of a sodium-chloride-doped polycrystalline specimen in a flexural mode, as determined by Kuroiwa (1964) for four different vibration frequencies. It is immediately clear that there are three different loss processes which become dominant in different temperature ranges. The loss peak, which varies with frequency over the temperature range −50 to −80 °C for the vibrations studied, is simply the intrinsic

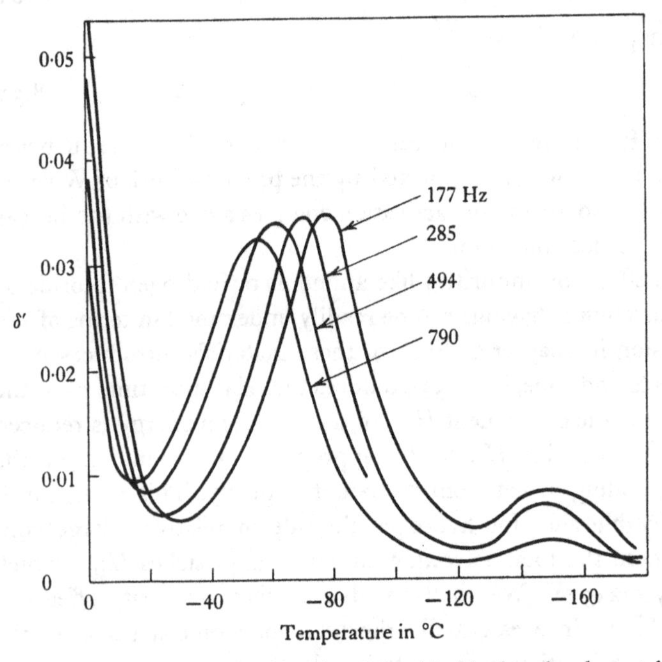

Fig. 8.7. The internal damping of sodium-chloride-doped polycrystalline ice samples for different resonant frequencies (after Kuroiwa, 1964).

molecular relaxation which we have already discussed. It is the only peak observed in single crystals and, since a polycrystalline sample presents a range of crystallite orientations, the value of δ_{max} is a weighted average of the values in table 8.3.

The loss above −30 °C and the peak near −150 °C occur only in polycrystalline samples and must therefore be associated in some way with the grain boundaries. Above −30 °C there is no sign of a peak but simply a region of sharply increasing loss as the temperature approaches the melting point. The curves do, however, shift systematically with frequency and from this shift an apparent activation energy can be determined, ranging from 1·3 eV for impure ice to 2·6 eV for pure ice. The origin of the grain boundary damping is fairly clearly in some sort of quasi-viscous flow which is accentuated if there is segregation of impurity at the boundary leading to boundary melting at temperatures below 0 °C. The measurements have not yet been interpreted in terms of any detailed physical mechanism.

Finally we have the peak near −150 °C, which is not present in pure monocrystalline or polycrystalline ice and is therefore associated with some impurity, in this case sodium chloride. The temperature of the peak does not shift with frequency, though its magnitude does change. Similar effects are found with other impurities but their detailed interpretation is again not clear.

8.4. Dislocations, creep and plastic flow

For any material, Hooke's law, even with the generalization of complex elastic moduli, is only an approximation, valid in general if the stresses involved are not too large and are not applied for too long a time. If the stresses are large, the specimen may suffer a permanent deformation or even fracture, while stresses well below these values, if applied for many hours, may lead to creep. These phenomena are of considerable practical interest, not only for their implication for the design of structures to be supported upon ice foundations, but also for an understanding of the flow of glaciers. We shall rather be concerned, however, with the nature of the flow processes taking place in single crystals.

From our discussion of the structure of ice in chapter 2 and the picture given in fig. 8.8, an ice crystal can be regarded as a stack of crinkled molecular sheets, each having hexagonal symmetry and lying perpendicular to the *c*-axis. Half of the molecules in a sheet have a single bond to the sheet below and half to the sheet above, so that it is not surprising that the principal type of plastic deformation encountered is a glide of these sheets over one another, which we may call basal glide. It should be emphasized, however, that the bonds parallel to the *c*-axis are just as strong as those in other directions, unlike the case of a graphite crystal for example, so that the phenomenon is simply a preference for glide on a particular set of planes rather like that found for close-packed {111} planes in face-centred cubic crystals.

This mode of deformation has been known for a long time and the experimental evidence prior to 1958 has been reviewed by Glen (1958). Experiments on single crystals are obviously of great help because there must be many extraneous effects in polycrystal-

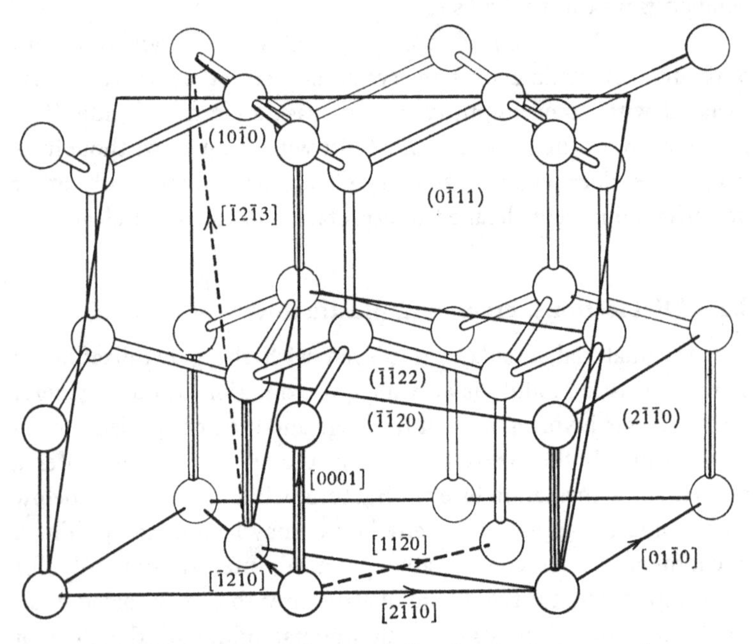

Fig. 8.8. Prominent planes and directions in the ice crystal.

line specimens. Single crystals can be grown in the laboratory but it is difficult to achieve uniform properties and freedom from strain in large crystals; however, the Mendenhall Glacier in Alaska furnishes single crystals up to several kilogrammes in weight, presumably by a centuries-long process of strain annealing, which have been found to be of exceptional purity and mechanical perfection, and these have been used in several laboratories for deformation studies.

Experiments are generally carried out in one of two ways. Either a constant stress is applied to the crystal and its strain is measured as a function of time, giving a creep curve, or else a testing machine is used to increase the strain at a constant rate and the associated stress is measured. In either case the strain may be an elongation, a compression or some sort of bending, different advantages in experiment or interpretation applying for different modes.

The shape of a typical creep curve, obtained in this case by bending a single crystal rod with the crystal *c*-axis normal to both

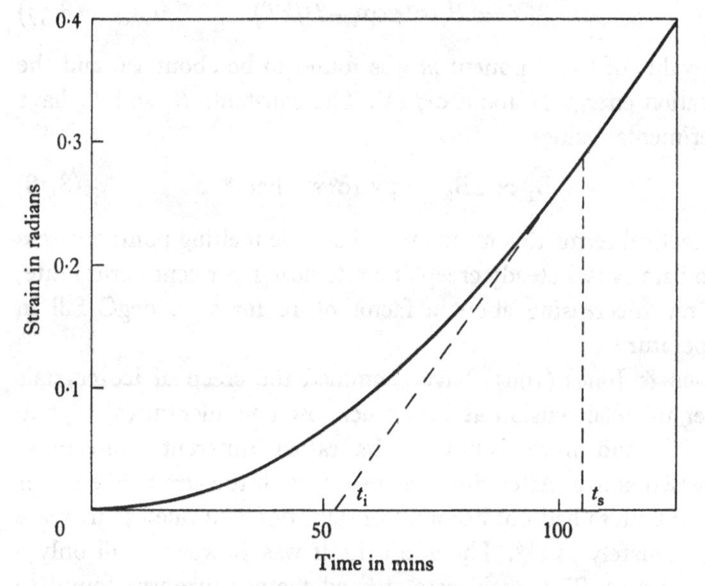

Fig. 8.9. A typical creep curve for an ice single crystal slipping on basal planes (after Higashi *et al.* 1965).

the length of the rod and axis of bending so that deformation involves basal slip, is shown in fig. 8.9, due to Higashi *et al.* (1965). These workers found that exactly similar curves, differing only in scale, resulted for tests at temperatures from −4·8 to −40 °C and stresses from 18 to 60 bars, using in all cases natural glacier crystals. Curves of a generally similar shape were found by Steinemann (1954) at a temperature of −2·3 °C for stresses in the range 0·45–2·2 bars using laboratory crystals and by Jellinek & Brill (1956) at −5 °C for a stress of 0·5 bars using glacier crystals.

The features of all these curves are an initial period, shown as

t_s in fig. 8.9, in which the strain varies more or less quadratically with time, followed by a regime of steady-state flow with strain increasing linearly with time. Higashi *et al.* find that the incubation time t_1, which is just half t_s if the first part of the curve is parabolic, is related to stress σ and to temperature by

$$t_i^{-1} = B_1\sigma^m \exp(-H/kT), \tag{8.36}$$

while the steady-state creep rate $\dot{\epsilon}$ is given by

$$\dot{\epsilon} = B_2\sigma^m \exp(-H/kT). \tag{8.37}$$

The value of the exponent m was found to be about 1·6 and the activation energy H about 0·7 eV. The constants B_1 and B_2 have experimental values

$$B_1 \simeq 2B_2 \simeq 4\times 10^6 \text{ s}^{-1} \text{ bar}^{-m}. \tag{8.38}$$

In practical terms this means that, near the melting point, a stress of 50 bars causes steady creep at a rate near 1 per cent per minute, this rate decreasing about a factor of 10 for a 10 degC fall in temperature.

Glen & Jones (1967) have examined the creep of ice crystals under uniaxial tension at very much lower temperatures, -50 to -70 °C, and have found results rather different from those discussed above. After times as long as 50 h the strain rate under constant stress had not become constant but continued to increase approximately as $t^{1.5}$. The strain itself was, however, still only a few per cent. The strain rate at fixed temperature was found to behave as in (8.37) but the exponent had the much larger value $m = 4\pm 1$. The reasons for these differences are not clear but may be due to differences in temperature, deformation mode or crystal perfection in the specimens used.

Experiments of the second type, in which the strain is increased linearly with time, give the sort of results shown in fig. 8.10, from work by Higashi *et al.* (1964), this time for glacier crystals in tension with the basal plane at 45° to the strain axis; closely similar results have been found by Ready & Kingery (1964). There is a clear yield point at a strain between 0·5 and 1 per cent, after which the stress decreases markedly. This is a familiar phenomenon in metals but, in distinction from the metallic case, the stress in ice remains low above the yield point, with none of the work-hardening found in metals. The stress at the yield point was found by Higashi *et al.* to be related to strain rate and temperature by

$$\sigma_{\max} = B_3 \dot{\epsilon}^{1/m} \exp(H'/kT), \tag{8.39}$$

with $m = 1{\cdot}53$ and $H' = 0{\cdot}45$ eV. This equation is obviously similar to (8.37) and, when written in the same form, the exponent m and the activation energy $mH' = 0{\cdot}7$ eV are very close to the previous values.

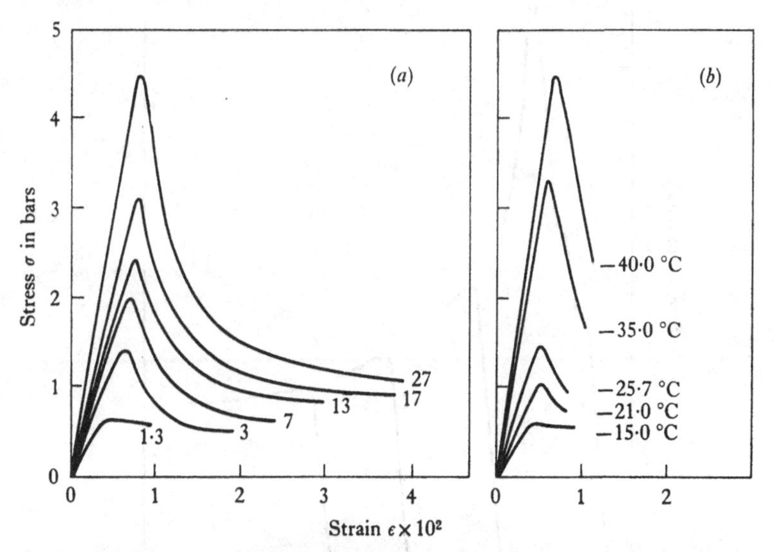

Fig. 8.10. Stress–strain curves for ice single crystals deformed by basal shear at constant strain rate. In (*a*) the temperature is −15 °C and the strain rate is given as a parameter in units of 10^{-7} s^{-1}, while in (*b*) the strain rate is $1{\cdot}3 \times 10^{-7}$ s^{-1} and the temperature is varied (Higashi *et al.* 1964).

If the stress–strain behaviour after flow has commenced is examined, then we find a law of the form (8.37), the exponent m determined by Ready and Kingery varying from about 2·5, for crystals strained by less than about 10 per cent, to 1·5 for heavily strained crystals. This latter result agrees well with the creep experiments discussed above, while the measured activation energy, $H = 0{\cdot}62 \pm 0{\cdot}06$ eV, is only slightly smaller.

If, after plastic yielding has commenced, the experiment is stopped and then started again, the sort of results shown in fig. 8.11

are found. The ice crystal does not recover its original hardness but begins to flow at a much lower stress and in a way which is just a continuation of the original curve. Again, if an ice crystal is held with a static strain ϵ and the rate of decay of the stress is followed, then we find (Ready & Kingery, 1964)

$$\dot{\sigma} = -B_4\sigma^m, \tag{8.40}$$

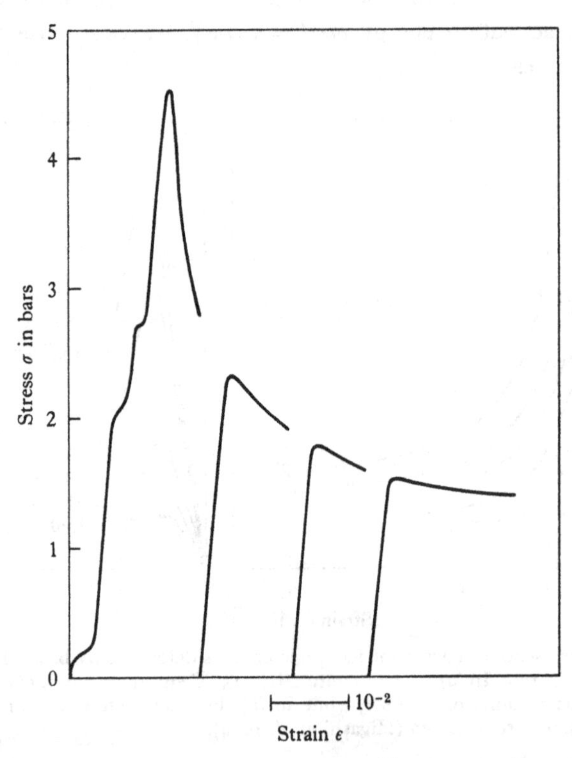

Fig. 8.11. Interrupted stress–strain curve. Temperature is −16 °C and strain rate 2×10^{-7} s^{-1} (Higashi *et al.* 1964).

where the exponent m for small strain is $2{\cdot}6 \pm 0{\cdot}3$, suggesting a connexion with the stress–strain behaviour for similar small strains.

A description of these phenomena in terms of processes occurring at the molecular level in ice is simple in outline, though many details are not yet properly understood. The treatment is similar to that for metals (Cottrell, 1953) or to the more closely analogous case of ionic crystals (Johnston, 1962). When an ice crystal is stressed, then the total observed strain rate $\dot{\epsilon}$ is the sum of an elastic part $\dot{\epsilon}_e$ and a plastic part $\dot{\epsilon}_p$:

$$\dot{\epsilon} = \dot{\epsilon}_e + \dot{\epsilon}_p. \tag{8.41}$$

The elastic part includes deformation of the ice crystal, its mounting and the testing machine as a whole while, in a well-designed experiment, $\dot{\epsilon}_p$ should be wholly confined to the crystal under test. The actual plastic deformation is caused by motion of dislocations in the basal plane, a subject to which we return presently. If, then, n is the concentration, v the velocity and b the magnitude of the Burgers vector of these dislocations, we can write

$$\dot{\epsilon}_p = nbv. \tag{8.42}$$

The dislocation velocity, once the stress σ is large enough to cause motion, can be assumed to follow a law of the form

$$v = A\sigma^m, \tag{8.43}$$

where the exponent m is characteristic of the material involved. This motion will presumably involve some sort of energy barrier H, so that we expect

$$A = A_0 \exp(-H/kT). \tag{8.44}$$

As the dislocations move, they generate new dislocations and the simplest reasonable assumption is that total dislocation density is proportional to strain

$$n = B\epsilon_p. \tag{8.45}$$

We must, however, allow for the presence of some initial dislocations and this is most easily done by assuming a small initial value of ϵ_p at $t = 0$.

Equations (8·42)–(8·45) can now be combined to give

$$\dot{\epsilon}_p = A_0 Bb\epsilon_p \sigma^m \exp(-H/kT), \tag{8.46}$$

which agrees well with the experimental observations for the part of the curve where $\epsilon_p \gg \epsilon_e$. Inclusion of the elastic term in (8.41) requires a numerical treatment but the general behaviour is clear. For small strains, $\epsilon_e > \epsilon_p$ and the stress–strain curve rises linearly. However, dislocations multiply rapidly at higher stresses and soon ϵ_p becomes dominant. The numerical calculations show that a sharply peaked initial stress behaviour is associated with a small value of m, which accords with the value near 2 found for ice.

Some of these microscopic equations have been checked directly by observation of the motion of small-angle grain boundaries under stress (Higashi & Sakai, 1961*a*, *b*). The behaviour followed (8.43) and (8.44) with $m \simeq 1$ and H in the range 0·5–0·75 eV, which is in reasonably satisfactory agreement with values derived from macroscopic experiments.

Several points in this general treatment require further comment. In the first place we have neglected interaction between dislocations, except for the multiplication equation (8.45). One might have expected A in (8.43) to depend on the dislocation density n as in metals, where such interaction impedes dislocation motion and leads to work-hardening. This does not occur in ice. Secondly, if we consider a normal creep experiment with σ constant, then (8.46) implies an exponential increase of strain with time. This does not occur and $\dot{\epsilon}_p$ tends to a constant. The probable explanation is that, when the dislocation density becomes high, dislocations can climb by a diffusion mechanism (Weertman, 1957) to annihilate each other after a limited amount of motion, thus maintaining n constant.

To look more closely at the dislocations involved in basal glide, we must return to the crystal structure of ice, as shown in fig. 8.8. The dislocations with lowest energy of formation will be those with the smallest Burgers vector, which is clearly $(a/3)\ \langle 11\bar{2}0\rangle$; all other dislocation types have much higher energy. Dislocations in ice have been studied by means of X-ray diffraction topography, a technique which allows them to be photographed directly using X-rays Bragg-reflected from a particular set of crystal planes, by Webb & Hayes (1967). A typical photograph is shown in plate 3. The crystals used were planar dendrites, grown in supercooled water, and the initial specimens were found to be almost entirely dislocation-free. Gentle straining, however, developed extensive dislocation networks, as shown, and the topographic technique established that the dislocations lie predominantly in the (0001) plane, run in $\langle 11\bar{2}0\rangle$ directions and have Burgers vectors $(a/3)\ \langle 11\bar{2}0\rangle$ as expected. The association between dislocation direction and Burgers vector was such that the screw orientation, with Burgers vector parallel to dislocation line, was most common. These are the dislocations involved in basal glide. Other sorts of dislocation can, of course, occur but are much rarer.

Another method of studying dislocations is by etching the crystal surface. Molecules close to the dislocation are under elastic stress and are hence more easily removed than molecules in a normal crystal, giving rise to an etch pit. A suitable etchant, developed by Higuchi (1958), consists of 1–5 per cent of polyvinyl formal (Formvar) dissolved in ethylene dichloride and applied to the polished crystal surface. The solvent, in evaporating, acts as a powerful surface etch and the polymer, as well as moderating its action, produces a surface replica film which can be examined by optical or electron microscopy. Examination of unstrained ice crystals by this means by Bryant & Mason (1960) indicated an etch pit density on the basal surface, and hence presumably a density of dislocations cutting this surface, of order 10^5 cm^{-2}. This is, of course, not a general figure, but indicates the degree of perfection of the particular crystal studied, this density being comparable to that found in silicon or germanium crystals of moderate quality.

The way in which dislocations move in an ice crystal and the process determining the activation energy H in (8.44) also merit further study. Motion of a dislocation over one intermolecular distance normal to itself involves a change in bonding pattern all along the dislocation line. The energy involved in the elementary process of breaking a bond is not large, only about 0·3 eV in unstrained crystal and certainly less near the dislocation, so this cannot account for the observed activation energy of 0·6–0·7 eV. However, as Bjerrum (1952) pointed out and Glen (1968) has recently considered in detail, simple transfer of bonds without attention to proton configuration would result in half the bonds forming L- or D-defects, leading to a prohibitively large energy penalty for motion. Before the dislocation can move, therefore, it is necessary for the proton configuration to change so that bonds can be properly made. This can only occur, in pure ice, through the migration of thermally generated D- and L-defects and the activation energy associated with the formation and motion of these is, as we saw in chapter 7, about $H = \frac{1}{2}H^{\mathrm{f}}_{\mathrm{DL}} + H^{\mathrm{t}}_{\mathrm{DL}} = 0{\cdot}58$ eV, in moderate agreement with the value deduced from the present mechanical measurements. The matter is clinched by the work of Jones (1967), who found that a few parts per million of hydrogen fluoride in the ice crystal increases the creep rate by an order a magnitude at −70 °C, as would be expected if the diffusion of L-defects controls dislocation velocity.

The mechanism of bond reorientation is not necessarily simple and may involve molecular orientation in the strain field ahead of the moving dislocation, the coupling mechanism being the same as that which we discussed for elastic relaxation effects. Such a reorientation contributes a viscous drag to dislocation motion, as proposed by Schoeck (1956) and elaborated by Eshelby (1961), quite independently of the orientation effects at the dislocation line itself, though the two are coupled through sharing the same relaxation mechanism and therefore show the same activation energy.

It is interesting at this point to compare the plastic behaviour of ice crystals with that of other tetrahedrally bonded crystals such as silicon or germanium, a comparison we have already found helpful in considering thermal properties. Very nearly perfect crystals of these semiconductors can be grown and their deformation has been studied by Patel & Chaudhuri (1963) at temperatures between 0·5 and 0·8 of their absolute melting points, corresponding roughly to the experimental conditions for ice. Since these materials have diamond cubic structure, the planes on which easy slip can occur are the {111} planes, corresponding to the basal plane of ice in structure and bonding. There will therefore always be more than one slip system active during deformation, giving rather more complex behaviour than that of ice.

Silicon at 900 °C and germanium at 500 °C both show very pronounced yield points, when the initial dislocation density is low, and these correspond very closely in shape and position to those found for ice. Since a pronounced yield point is associated in the theory with a small value of the exponent m in the dislocation velocity relation (8.43), it is satisfying that direct observation of dislocation motion in these materials (Chaudhuri *et al.* 1962) gives m from 1·3 to 1·9 in close similarity to the value $m \simeq 1$ found by Higashi & Sakai (1961*b*) for ice or the value between 1·5 and 2·5 deduced from macroscopic experiments. These values are in sharp distinction to exponents as high as 16·5 found by Johnston (1962) for lithium fluoride and up to 44 (Chaudhuri *et al.* 1962) for silicon-iron. We can therefore conclude that the low exponent and resulting sharp yield point are direct consequences of the tetrahedral bonding in ice.

One further point is worth comment. This is that, for certain deformation modes in both germanium and silicon, there is no work-hardening up to strains of several per cent, in agreement with the much more extreme results for ice. For other orientations, however, work-hardening does occur, possibly because of interaction between the different slip systems.

Another apparent anomaly in the plastic behaviour of ice is the absence of any preferred glide direction in the basal plane. Metals glide preferentially in the most closely packed direction of the most closely packed plane and we might expect similar preference for glide in the $\langle 11\bar{2}0\rangle$ directions in ice, since the necessary dislocations exist. Instead of his, the glide is nearly isotropic in the (0001) plane (Glen & Perutz, 1954).

This paradox was resolved by Kamb (1961), who showed that it is a consequence of the hexagonal symmetry and stress dependence of flow in ice. From (8.37) or (8.46) the flow law is

$$\dot{\epsilon} = A\sigma^m \tag{8.47}$$

and, whilst this should give isotropic behaviour for $m = 1$, this seems unlikely for the experimental $m \simeq 1{\cdot}6$. Kamb showed, however, using (8.47) for the resolved stress along each glide direction, that the glide is also parallel to the stress if $m = 3$. For $m > 3$, which is the case for metals near room temperature, the glide tends more and more closely towards one of the preferred directions. For $1 < m < 3$ we reach the rather surprising conclusion that the resultant glide tends slightly away from, rather than towards, the nearest glide direction. The deviation from the applied stress direction is, however, small in this case, amounting to less than 2°, so that glide is apparently istropic in the (0001) plane, as observed for ice.

So far we have considered only slip parallel to the basal plane in ice and, though this is a very strongly preferred deformation mode, it is not the only one which can occur. Experiments to search for activity on other slip planes require that the resolved shear stress on basal planes be zero, which means that, in a tensile or compressive test, the *c*-axis must be either parallel or perpendicular to the stress axis for the specimen. Muguruma *et al.* (1966) used the former orientation in tension and observed the formation of plate-shaped voids in the basal plane whose origin they explained tentatively in terms of non-basal slip. It is interesting to note that the stress–strain curve in this case, as well as having a yield point, gave evidence of work-hardening for higher strains, in contradistinction to the case of basal glide. Slightly more direct evidence of non-basal glide comes from an electron microscope study made by Muguruma & Higashi (1963*a*, *b*) of etch channels on (0001) planes of crystals so strained as to produce this type of slip. Etching showed channels running in both $\langle 11\bar{2}0\rangle$ and $\langle 10\bar{1}0\rangle$ directions and, if these are interpreted as tracks left by moving dislocations, then these dislocations must have shed vacancies along the track as they moved, the vacancies leading perhaps to a precipitation of impurities along the track. Slip systems capable of producing this result are $\{10\bar{1}0\}\langle\bar{1}\bar{1}23\rangle$, $\{10\bar{1}0\}\langle 11\bar{2}0\rangle$ and $\{11\bar{2}2\}\langle\bar{1}\bar{1}23\rangle$. The crystal structure shown in fig. 8.8 would also lead one to expect a slip system of the type $\{10\bar{1}1\}\langle\bar{1}2\bar{1}0\rangle$, since this is a quite closely packed direction on a well-populated crystal plane but, since motion of such dislocations does not produce a vacancy trail, their path would not be detected by the etching procedure.

Finally, let us consider briefly the deformation of polycrystalline ice, since this is a phenomenon of practical importance. Many experimental studies have naturally been of an engineering or glaciological nature and these are outside the scope of this book. The reader who is interested in these topics should consult the variety of articles and photographs published in the *Journal of Glaciology*, the mathematical treatment of glacier flow given by Nye (1951), the recent reviews by Pounder (1965) and Glen (1967) and the important collections of papers edited by Kingery (1963) and Ôura (1967).

The phenomenon of relegation, particularly in the form in which a loaded wire cuts through a block of ice and the ice then heals up after it, has also excited attention at various times during the past

100 years. The primary mechanism involved is undoubtedly pressure melting with subsequent refreezing, the latent heat involved being transported through the moving body, but many of the details remain obscure. A recent discussion with references to earlier work is given by Nye (1967).

The creep of polycrystalline ice has been studied by Glen (1955) over a stress range from 1 to 10 bars at temperatures 0 to -13 °C and by Jellinek & Brill (1956) from 0·3 to 2·5 bars and -5 to -15 °C. The behaviour is rather different from that of single crystals in that the creep rate is initially high and then falls with increasing strain, approximately in the form

$$\epsilon \simeq At^{\frac{1}{3}} + Bt, \tag{8.48}$$

where A and B are constants. This behaviour only applies for fairly small strains ($\epsilon < 0{\cdot}1$); at larger strains the specimen may recrystallize with a preferred texture so that slip can occur simultaneously upon all crystallites, approximately in the basal plane.

If Glen's results are analysed by simply fitting a straight line to the limiting strain rate $\dot{\epsilon}_s$, then this is related to stress by a law similar to that for single-crystal ice (8.48) but with a value near 3·2 for the exponent m. If, instead, $\dot{\epsilon}_s$ is taken to be the value of B in (8.48), the same law results but with m about 4·2. The temperature range of the experiments is not large but, if a temperature factor $\exp(-H/kT)$ is assumed, then Glen's results give $H \simeq 1{\cdot}4$ eV, while Jellinek and Brill find $H \simeq 0{\cdot}7$ eV. The explanation of this discrepancy is not clear. It may be due in part to the different stress and temperature ranges involved or perhaps the temperature variation does not have the simple form assumed because of the complicated intergranular distortions involved.

CHAPTER 9

ELECTRICAL PROPERTIES

The electrical properties of ice have been studied, in recent years, more extensively than any other of its attributes. One reason is that a large variety of phenomena can be conveniently classified under this heading, while another is that, because many electrical properties are very sensitive to the purity of the crystal, the number of possible results is greatly multiplied and a detailed interpretation, though necessarily complicated, gives a considerable amount of insight into the molecular processes involved. Before we examine any topic in detail, therefore, let us make a brief survey of the field to be discussed to see what the phenomena are and, in general terms, how they can be interpreted.

9.1. Introductory survey

First consider the dielectric constant of pure ice as a function of frequency, as shown schematically in fig. 9.1. The static dielectric constant ϵ_s is very large, close to 100, and the mechanism responsible for it may reasonably be identified as the orientation of molecular dipoles preferentially in the direction of the field. The value of ϵ_s does not vary greatly with a small amount of impurity in the specimen, in agreement with this interpretation. As the frequency is increased through the kilohertz range, ϵ falls dramatically to a high-frequency value $\epsilon_\infty = 3{\cdot}2$ which is maintained through the microwave region (Lamb & Turney, 1949). The frequency at which the fall in ϵ occurs can be associated with the inverse relaxation time for reorientation of molecular dipoles and depends strongly on temperature and specimen purity, which is what we might expect, since we saw in our discussion of molecular reorientation in chapter 8 that reorientation can only occur through the motion of point defects through the lattice. The phenomena leading to the observed value of ϵ_∞ are twofold: the molecules in the ice structure can rotate slightly without causing any bond rearrangement and the bond angles can distort to a small extent, and to this must be added the polarizability of the molecular electron cloud.

The frequencies of the principal lattice vibrations in ice lie, as we saw in chapter 6, in the far infrared, while the librations and molecular distortions have characteristic frequencies in the near infrared. Across these regions, therefore, the dielectric behaviour is again complex but, by the time the visible region is reached, all

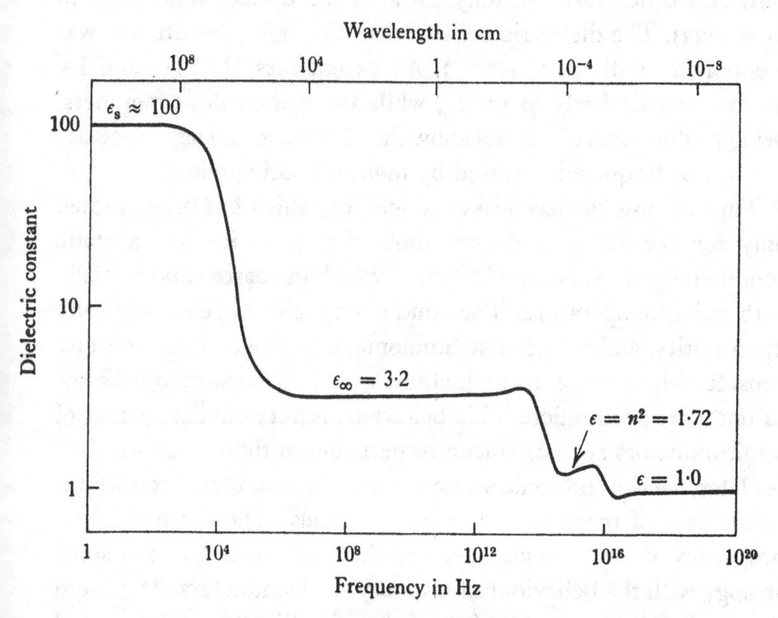

Fig. 9.1. Schematic representation of the behaviour of the dielectric constant of ice, ϵ, as a function of frequency at a temperature near -10 °C.

molecular absorption bands are past and the dielectric constant arises simply from electronic polarization. It therefore has the optical value $\epsilon \simeq 1{\cdot}72$, which is just the square of the refractive index n, which ranges from 1·306 for red light to 1·318 for violet, with a small amount of crystalline anisotropy.

The visible region of the spectrum is an almost transparent window for water in all its phases, which is one reason for the development of an organ sensitive in this wavelength range for terrestrial life, which originated largely in the oceans. Ultraviolet absorption for water in all its phases begins below about 1800 Å, so that in this region the dielectric behaviour is again complex and ϵ ultimately falls to its X-ray value, close to unity.

It is worth mentioning, in passing, the dielectric behaviour of water and of the high-pressure ices. Liquid water has a static dielectric constant of 78·3 at 25 °C and the dispersion due to elimination of molecular rotation does not occur until the microwave region ($\lambda \sim 1$ cm). The ϵ_∞ plateau region is thus hardly seen and ϵ decreases steadily towards the optical value of 1·78 ($n = 1{\cdot}33$). The dielectric behaviour of the high-pressure ices was mentioned briefly in chapter 3. All except Ices II, VIII and IX behave very similarly to Ice I_h, while these anomalous members, being proton ordered, do not show the dispersion and large increase in ϵ at low frequencies caused by molecular orientation.

Turning now to electrical conductivity, which has been studied only for Ice I_h, experiments show that pure ice has a static conductivity of order $10^{-9}\ \Omega^{-1}\ \text{cm}^{-1}$ which increases exponentially with rising temperature. The conductivity also depends critically upon purity, and solutes like ammonia or hydrogen fluoride cause considerable increase in conductivity even in concentrations as low as one part per million. This behaviour is very similar to that of semiconductors such as silicon or germanium though, as we shall see later, ice is a 'protonic semiconductor' rather than electronic as in the case of these more familiar materials. The thermoelectric properties of ice have also been studied and again there is some analogy with the behaviour of ordinary semiconductors. Measured thermoelectric powers are of the order of millivolts per degree and depend strongly upon purity, in contrast to thermoelectric powers of tens of microvolts per degree characteristic of metals.

There are many other electrical phenomena associated with ice, such as frictional charging, which we do not have space to consider and others, such as Hall effect and the possibility of a ferroelectric transition, which must be mentioned only briefly. In the following sections we shall discuss in more detail the main features of the electrical behaviour outlined above.

9.2. Static dielectric constant

The dielectric properties of a material are properly specified by a symmetric second-rank tensor relating the three components of the electrical displacement vector **D** to those of the field **E**. By choosing axes naturally related to the crystal structure the six independent components of this tensor can be reduced to three and, taking account of the hexagonal symmetry of the ice crystal, only two independent components remain. These are the relative permittivities parallel and perpendicular to the unique *c*-axis direction and we shall denote them by $\epsilon_{\parallel}$ and $\epsilon_{\perp}$. We shall discuss the experimental determination of these quantities when we come to consider dielectric relaxation, since some difficulties are involved. For the present we simply note the results which are shown in fig. 9.2. The often-quoted careful measurements of Auty & Cole (1952) were made with polycrystalline samples and removed many of the uncertainties in earlier work. They represent, however, a weighted mean of the values of $\epsilon_{\parallel}$ and $\epsilon_{\perp}$. Humbel *et al.* (1953*a*), on the other hand, used single crystals and found $\epsilon_{\parallel} > \epsilon_{\perp}$, the anisotropy being about 14 per cent. Since Humbel *et al.* only claim an absolute accuracy of about 4 per cent, their measurements are in acceptable agreement with those of Auty and Cole. The static dielectric constant of D_2O ice was found by Auty and Cole to be essentially equal to that of H_2O ice from 0 to -33 °C.

Calculation of the static dielectric constant is in principle a simple matter, though in practice it presents many difficulties. In an ice crystal, as we have seen, the molecular dipoles are directed at random, subject only to the Bernal–Fowler rules. This means that the energy of all such possible configurations is the same, or at least that the energy differences are small compared with kT at liquid-air temperature. When an electric field **E** is applied, configurations having a positive net dipole in the direction of the field are energetically favoured and the equilibrium polarization $(\epsilon_s - 1)\,\mathbf{E}$ can be calculated by the methods of statistical mechanics. A review of the various methods and approximations used in this general development can be found in the book by Fröhlich (1949, chapter 2).

Suppose we have a macroscopic system such as an ice crystal, and let us concentrate our attention upon a small spherical region of volume V inside that system which is large enough to have the same dielectric properties as the macroscopic body. Let the internal co-ordinates x_i of the material in this sphere be denoted collectively by X, then a particular configuration will have, in

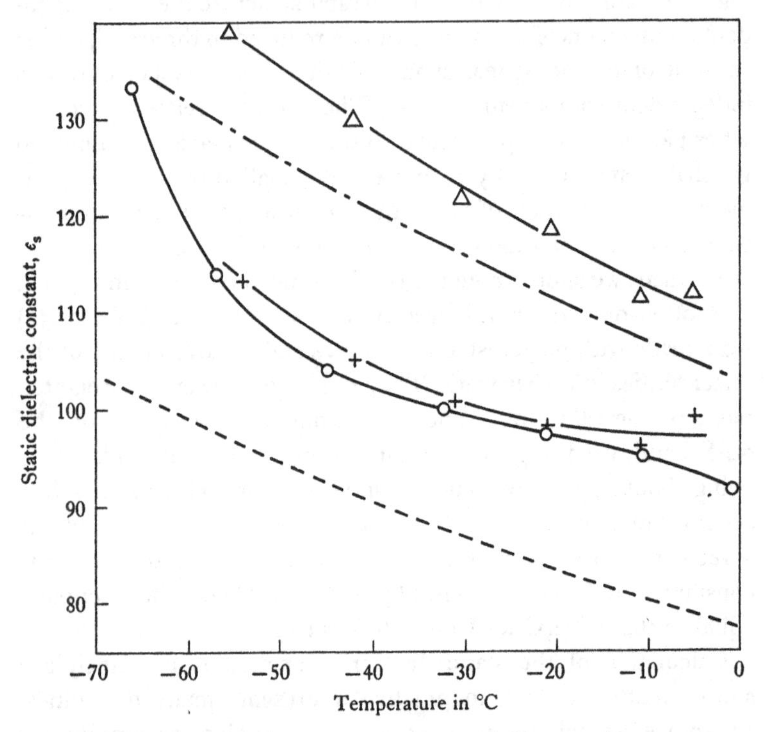

Fig. 9.2. Variation of the static dielectric constant of ice, ϵ_s, with temperature according to theory and calculation. ○, Auty & Cole (1952) polycrystalline; △, Humbel *et al.* (1953*a*) $||_c$; +, Humbel *et al.* (1953*a*) $\perp_c$; - - - - -, Powles (1952), twenty-four molecules, equal probability; — - —, Powles (1952), seventeen molecules, weighted probability.

general, an energy $U(X)$ and an electric moment $\mathbf{M}(X)$. If an electric field $\mathbf{E}$ is applied to the crystal, then, from electrostatic considerations, the homogeneous field acting on the spherical region, considered as exactly filling a spherical cavity in the bulk material, is

$$\mathbf{E}' = \left(\frac{3\epsilon_s}{2\epsilon_s + 1}\right)\mathbf{E}, \tag{9.1}$$

where ϵ_s is the static dielectric constant of the bulk medium. In the presence of this field, the energy $U(X,E)$ of the spherical region is thus

$$U(X,E) = U(X) - \left(\frac{3\epsilon_s}{2\epsilon_s+1}\right) M(X)E\cos\theta, \tag{9.2}$$

where θ is the angle between $\mathbf{M}(X)$ and $\mathbf{E}$. The moment induced by the field $\mathbf{E}$ is thus

$$M_E = \frac{\int M(X)\cos\theta\exp[-U(X,E)/kT]\,dX}{\int\exp[-U(X,E)/kT]\,dX}, \tag{9.3}$$

which, using (9.2) in the limit of small E and remembering that $M_{E=0} = 0$, gives

$$M_E = \left(\frac{3\epsilon_s}{2\epsilon_s+1}\right)\frac{E}{kT}\frac{\int M^2(X)\cos^2\theta\exp[-U(X)/kT]\,dX}{\int\exp[-U(X)/kT]\,dX} \tag{9.4}$$

We have, in (9.1), implicitly assumed the material to be isotropic, which is not a bad approximation for an ice crystal, so that in the integral we may replace $\cos^2\theta$ by its average value $\frac{1}{3}$. Then, recalling that the dielectric constant is related to the induced moment per unit volume by

$$\epsilon_s - 1 = 4\pi M/VE, \tag{9.5}$$

(9.4) gives
$$\epsilon_s - 1 = \frac{4\pi}{3V}\left(\frac{3\epsilon_s}{2\epsilon_s+1}\right)\frac{\langle M^2\rangle}{kT}. \tag{9.6}$$

$$\langle M^2\rangle = \int M^2(X)\exp[-U(X)/kT]\,dX \Big/ \int\exp[-U(X)/kT]\,dX, \tag{9.7}$$

which is the mean square of the spontaneous polarization of the spherical region when embedded in a large specimen of the same material.

$\langle M^2\rangle$ can be evaluated by considering $\mathbf{M}(X)$ as a sum over the contributions $\mathbf{m}(x_j)$ for the individual internal co-ordinates. There is a further simplification if all the x_j are divided into small groups representing the co-ordinates of a single molecule, for then the contribution from each molecule is the same and we can write

$$\langle M^2 \rangle = N \langle \mathbf{m} . \mathbf{m}^* \rangle, \tag{9.8}$$

where N is the number of molecules in the spherical volume, $\mathbf{m}$ is the dipole moment of a single molecular unit, $\mathbf{m}^*$ is the average dipole moment of the whole spherical region embedded in its own medium when the given molecule is held so as to have a moment $\mathbf{m}$, and $\langle \mathbf{m} . \mathbf{m}^* \rangle$ is the average value of the product $\mathbf{m} . \mathbf{m}^*$ taking into account all possible configurations. (9.6) therefore becomes

$$\epsilon_s - 1 = \left(\frac{3\epsilon_s}{2\epsilon_s + 1} \right) \frac{4\pi N}{3V} \frac{\langle \mathbf{m} . \mathbf{m}^* \rangle}{kT}. \tag{9.9}$$

This very general result is due to Fröhlich and may be specialized, as we shall see presently, to give other well-known expressions. Before we do this, however, let us look briefly at the quantity $\mathbf{m}^*$. If there is no interaction between molecular dipoles then $\mathbf{m}^* = \mathbf{m}$ and indeed this can also be shown to hold if each molecule can be treated as a point dipole or as a uniformly polarized sphere. The deviation of $\langle \mathbf{m} . \mathbf{m}^* \rangle$ from $\mathbf{m}^2$ is thus a measure of the short-range interactions between molecules.

Again following Fröhlich, we can separate out the high-frequency polarizability ϵ_∞ in (9.9) to write it in the form

$$\epsilon_s - \epsilon_\infty = \left(\frac{3\epsilon_s}{2\epsilon_s + \epsilon_\infty} \right) \frac{4\pi N}{3V} \frac{\langle \mathbf{m} . \mathbf{m}^* \rangle}{kT}, \tag{9.10}$$

and, of course, this equation is further simplified if $\epsilon_s \gg \epsilon_\infty$, as is the case with ice.

The moments $\mathbf{m}$ and $\mathbf{m}^*$ appearing in (9.10) refer to molecules within the crystal and these differ from the molecular moments $\boldsymbol{\mu}$ determined in the vapour because of the polarizing effect of neighbouring molecules. If each molecule is treated as a uniform sphere, from the point of view of induced polarization, then (9.10) can be modified to

$$\epsilon_s - \epsilon_\infty = \left(\frac{3\epsilon_s}{2\epsilon_s + \epsilon_\infty} \right) \left(\frac{\epsilon_\infty + 2}{3} \right)^2 \frac{4\pi N}{3V} \frac{\langle \boldsymbol{\mu} . \boldsymbol{\mu}^* \rangle}{kT}. \tag{9.11}$$

This general form of Fröhlich's result can be simplified by making specific assumptions about the nature of the short-range intermolecular forces. If their range is taken to be small enough that only the z nearest neighbours of a given molecule are correlated with it in direction, then we can write

$$\langle \boldsymbol{\mu} . \boldsymbol{\mu}^* \rangle = \mu^2 g = \mu^2(1 + z\langle \cos\gamma \rangle), \tag{9.12}$$

where g is called the correlation coefficient and $\cos\gamma$ is the angle between the dipole of a molecule and that of one of its nearest neighbours. This assumption gives the form of result due to Kirkwood. If we go even further and assume there are no correlations at all, so that $\langle \cos\gamma \rangle = 0$, we have the result due to Onsager.

Onsager's equation applies properly to a dilute solution of molecules with permanent moments, while Kirkwood's can be usefully applied to liquids. Indeed, a simple correlation model for liquid water in which the tetrahedral co-ordination of nearest neighbours is preserved and more distant neighbours are neglected (Fröhlich, 1949, pp. 137–42) gives $\langle \cos\gamma \rangle = \frac{1}{3}$ or $g = 2{\cdot}33$ and predicts $\epsilon_s \simeq 63$ at 300 °K, in moderately good agreement with the experimental value 78. The T^{-1} variation predicted by (9.11) is also a good approximation. A more detailed calculation by Oster & Kirkwood (1943), based on a liquid model in which nearest neighbours are free to rotate about the bond joining them and using X-ray data to evaluate the nearest-neighbour distance and co-ordination number z, gives exact agreement with experiment at 25 °C and is within 12 per cent over the range 0–83 °C. The calculated correlation coefficient g ranges from 2·63 at 0 °C to 2·82 at 83 °C.

When we come to consider ice, a nearest-neighbour approximation is clearly inadequate and a much more careful calculation using the Fröhlich equation in the form (9.11) is needed. Such a calculation has been carried out by Powles (1952) under two different assumptions. In the first case he adopted Pauling's postulate that all molecular configurations in ice are equally probable and, using a cluster of twenty-four molecules as his spherical unit, calculated a value of ϵ_s of 77·3 at 0 °C with the T^{-1} temperature dependence shown in fig. 9.2. As an alternative model he made a correction for electrostatic interaction among molecules in the central region, thus modifying slightly the equal probability of the Pauling model. Because of the greater complexity, only seventeen molecules were considered in the group, and ϵ_s had the higher calculated value of 103 at 0 °C. Both these values are in reasonable agreement with experiment when the small size of the molecular cluster involved is taken into account—a group of at least twenty-one molecules is required if it is to exhibit the hexagonal rings characteristic of the ice structure. These calculations do not give any evidence about the anisotropy of ϵ_s, partly because of the small size of the molecular group considered and partly because of the assumptions implicit in the Fröhlich equation, though these can be extended (Powles, 1955). They do, however, bring out the fundamental role played by molecular correlations in determining the static dielectric constant of ice. In picturing the orientation of dipoles one must bear in mind, however, that the degree of orientation required in the structure to produce the observed dielectric effects is very small, a preferential orientation of only 1 in 10^7 being produced by a field of 1 V cm^{-1}.

In this discussion of dielectric constant the calculation was an equilibrium one and no account was taken of the way in which the equilibrium configuration is achieved. It was only necessary that some mechanism exist for the reorientation of molecular dipoles. In the next section we shall consider the mechanism of dielectric relaxation in detail and from this consideration will come an alternative method for calculating ϵ_s.

The static dielectric constants of the high-pressure ices have been measured as well and were given in table 3.3. Ices II, VIII and IX are proton ordered and have small static dielectric constants corresponding to ϵ_∞. The other ices have ϵ_s in the range 117–193 at -30 °C, which is roughly what might be expected since their densities under pressure are up to nearly twice that of Ice I_h at 1 bar. Hobbs *et al.* (1966) have given a theoretical treatment of these dielectric constants on the basis of significant structure theory, though the significant structures adopted, which consist of neighbouring antiparallel domains, seem to have little physical basis.

9.3. Dielectric relaxation

The dielectric relaxation of an ice crystal, involving as it does the rotation of molecular dipoles as described by certain internal co-ordinates, can be treated formally in very much the same way as in our discussion of mechanical relaxation in chapter 8. The dielectric problem was, however, treated long ago by Debye and is described in some detail, with specific reference to ice, in his book (Debye, 1929, chapter 6). Instead of following a development parallel to the mechanical case, therefore, we shall briefly remind ourselves of the fundamentals, see how the experimental facts fit into this simple framework, and then discuss the specifically electrical case in detail.

Considering for the present simply an isotropic material, the dielectric constant ϵ is defined by

$$D = \epsilon E \tag{9.13}$$

We know that if E is static then ϵ has the value ϵ_s, while for very rapidly varying fields $\epsilon = \epsilon_\infty$. Suppose now that the field E is simply switched on to a steady unit value at $t = 0$, so that it can be represented by the unit step function $u(t)$. The simplest reasonable assumption is that ϵ relaxes exponentially from ϵ_∞ to ϵ_s with a characteristic time τ so that

$$D(t) = [\epsilon_s + (\epsilon_\infty - \epsilon_s)\exp(-t/\tau)]\,u(t) = \epsilon(t)\,u(t). \tag{9.14}$$

Now what we really want is the behaviour of the dielectric under the influence of an alternating field of frequency ω so that

$$D(\omega)\exp(i\omega t) = \epsilon(\omega)E(\omega)\exp(i\omega t). \tag{9.15}$$

This suggests that we should take a Fourier transform of (9.14), making use of the fact that the transform of $u(t)$ is

$$u(\omega) = (2\pi)^{-\frac{1}{2}}(i\omega)^{-1},$$

so that

$$(2\pi)^{-\frac{1}{2}}(i\omega)^{-1}\epsilon(\omega) = (2\pi)^{-\frac{1}{2}}\int_{-\infty}^{\infty} D(t)\exp(-i\omega t)\,dt, \tag{9.16}$$

whence

$$\epsilon(\omega) = \epsilon_\infty + \frac{\epsilon_s - \epsilon_\infty}{1 + i\omega\tau}, \tag{9.17}$$

which is the form found by Debye. (Two comments are appropriate here. In the first place, some slight subtlety is needed in performing the Fourier transforms in order to ensure convergence; to do this we assume ω to have an infinitesimal imaginary part of appropriate sign. Secondly, the form (9.17) is that appropriate for an electric field represented by $E \exp(i\omega t)$; some authors use the opposite sign for ω.)

Equation (9.17) shows that the dielectric constant ϵ is in general complex. If we write

$$\epsilon(\omega) = \epsilon' + i\epsilon'', \tag{9.18}$$

then

$$\epsilon' = \epsilon_\infty + \frac{\epsilon_s - \epsilon_\infty}{1+\omega^2\tau^2}, \tag{9.19}$$

$$\epsilon'' = \frac{(\epsilon_s - \epsilon_\infty)\,\omega\tau}{1+\omega^2\tau^2}, \tag{9.20}$$

and the loss angle δ is given by

$$\tan\delta = \frac{\epsilon''}{\epsilon'} = \frac{(\epsilon_s - \epsilon_\infty)\,\omega\tau}{\epsilon_s + \epsilon_\infty\,\omega^2\tau^2}. \tag{9.21}$$

The form of these functions in the vicinity of the relaxation frequency is shown in fig. 9.3. This is the simplest form of dielectric relaxation behaviour and is generally termed Debye relaxation. The most obvious complication is that more than one relaxation process may be involved so that the assumption (9.14) is no longer adequate. Fortunately, however, the dielectric behaviour of ice is generally well described by this simple Debye theory.

There is an interesting and important consequence of the form of equations (9.19) and (9.20), which can be combined to give

$$\left[\epsilon' - \left(\frac{\epsilon_s + \epsilon_\infty}{2}\right)\right]^2 + (\epsilon'')^2 = \left(\frac{\epsilon_s - \epsilon_\infty}{2}\right)^2. \tag{9.22}$$

Thus, in the complex (ϵ', ϵ'')-plane, the locus of $\epsilon(\omega)$ is a semicircle which intersects the real axis (ϵ') at ϵ_s and ϵ_∞, its centre being on the real axis at $\frac{1}{2}(\epsilon_s + \epsilon_\infty)$. This type of dielectric plot, introduced by Cole & Cole (1941), is, as we shall see, of great help in interpreting experimental measurements. It also has a simple variant, for the case where there is a spread of relaxation times in the specimen, the centre of the circular plot being displaced slightly below the real axis.

A description of the detailed mechanism of dielectric relaxation in ice is complicated by the fact that our present understanding is derived from experiments not only with pure ice but also with ice containing a wide range of concentrations of proton-donor and proton-acceptor impurities. To simplify matters we will first discuss the experimental situation for pure ice, then put forward the detailed theory, and finally show that it can adequately account for the observations on doped crystals.

Measurements of dielectric dispersion in nominally pure ice have been made by many workers, including Smyth & Hitchcock

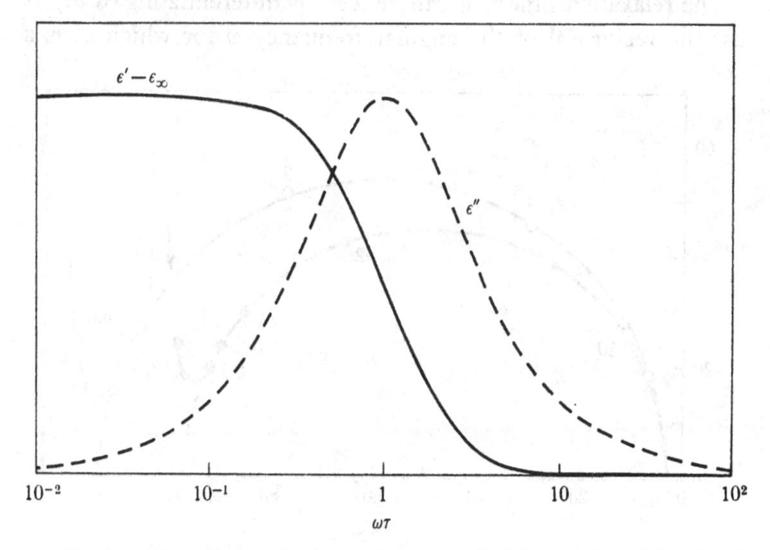

Fig. 9.3. The real part ϵ' and imaginary part ϵ'' of the complex dielectric constant near a region of Debye dispersion.

(1932), Auty & Cole (1952), Humbel *et al.* (1953*a, b*) and Steinemann (1957). These all differ in detail, though the later results are closely consistent.

Figure 9.4 shows the results of dielectric measurements on a single crystal of pure ice at −10·9 °C plotted on a Cole–Cole diagram. The anisotropy between $\epsilon_{\parallel}$ and $\epsilon_{\perp}$ at low frequencies is clearly visible and it can also be seen that this anisotropy vanishes at high frequencies. This diagram also shows the usefulness of the Cole–Cole plot. It is apparent that there is substantial deviation from a semicircular locus at low frequencies, the measured values of both ϵ' and ϵ'' becoming very large. This is an extraneous effect due to the finite static conductivity of the ice crystal and resulting from space-charge polarization effects at the electrodes (MacDonald, 1955) and depends upon temperature, sample dimensions and sample purity. At low temperatures and for very pure crystals this space-charge effect is not in evidence and the static dielectric constant can be measured directly; at higher temperatures the semicircular extrapolation gives the correct values for both ϵ_s and ϵ_∞.

The relaxation time τ, as can be seen by differentiating (9.20), is just the reciprocal of the angular frequency ω for which ϵ'' is a

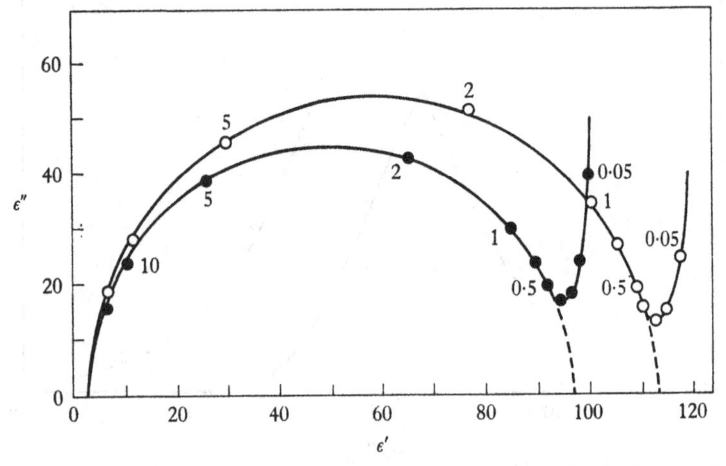

Fig. 9.4. Cole–Cole plot of the complex dielectric constant of a single crystal of ice at −10·9 °C. $\epsilon_\parallel$, ○—○; $\epsilon_\perp$, ●—●. The small numbers indicate measurement frequency in kHz. (From Humbel, Jona & Scherrer, 1953: *Helv. Phys. Acta.* **26**, 17–32, fig. 6; © Birkhäuser Verlag.)

maximum. Its measured value is about 10^{-4} s, at −10 °C, and the associated activation energy is about 0·58 eV (Auty & Cole, 1952; Steinemann, 1957).

The fact that the dielectric relaxation of ice is accurately described by a simple Debye curve implies that only a single mechanism is involved. Bjerrum (1951) discussed the possible reorientation mechanisms in ice—either ion-state motion or orientational defect motion—and concluded that the latter was the relevant process. As we shall see later, his conclusion is correct for pure ice, where the greater concentration of L- and D-defects more than makes up for their lower mobility, compared with ion states (see tables 7.1, 7.3), but for certain impurity contents it may be the ion states which dominate the relaxation.

More direct evidence for pure ice is, however, available from experiment. It is certainly possible for either ion states or orientational defects to cause molecular reorientation and the activation energy associated with the relaxation time τ must be that for the product of defect concentration and mobility. The measured value, 0·575 eV, is thus the sum of half the activation energy for formation of a defect pair together with the activation energy for defect diffusion. This is consistent with the values quoted for orientational defects in table 7.3 and is rather greater than the value for ion states derived from table 7.1. These activation energies are really enthalpies, however, and additional information can be derived from the dependence of the relaxation time on hydrostatic pressure. This has been investigated by Chan *et al.* (1965), who find that ϵ_s increases by about 3 per cent when the pressure is raised to 2000 bars, which is just about what would be expected from the increase in density. On the other hand, τ increases by about 30 per cent over this pressure range. This corresponds to a positive activation volume of $2{\cdot}9 \pm 0{\cdot}2$ cm^3 mole^{-1}. Now the formation of a pair of orientational defects certainly involves a positive volume change because of the repulsion between molecular vertices of like charge, while the creation of a pair of ion states leads to a decrease in volume because of the electrostatic field gradients around each ion. The observed activation volume is thus of the correct sign and is also of the correct magnitude for a process depending upon orientational defects, while the experimental evidence rules out quite conclusively a mechanism controlled by ion states.

One further piece of experimental evidence is required before we consider the theory in detail. This is the fact that pure ice crystals have a static conductivity which is by no means negligible, being of order $10^{-9}\ \Omega^{-1}\ \mathrm{cm}^{-1}$ at -10 °C, impure crystals generally having a much higher conductivity. We have already seen that this conductivity produces electrode polarization effects in dielectric measurements and similar effects may be troublesome in conductivity measurements. They can, however, be overcome by using special electrodes or high voltages.

Measurements have been made with platinum, platinum-black

or palladium foil electrodes, with evaporated gold electrodes, with 'sandwich' electrodes of ice heavily doped with hydrogen fluoride and even with ion-exchange membranes (Bradley, 1957; Jaccard, 1959; Eigen *et al.* 1964; Durand *et al.* 1967). Above about -5 °C the temperature dependence is anomalous, probably because of surface conductivity, but below this temperature the behaviour is consistent with an activation energy variously estimated to be between 0·40 and 0·61 eV. The absolute conductivity at -10 °C is in the range $(1{\cdot}0\text{–}1{\cdot}4) \times 10^{-9}\ \Omega^{-1}\ \text{cm}^{-1}$.

First we must ask whether this conductivity might be electronic, but the fact that the ultraviolet absorption threshold for ice is at about 1700 Å (Cassel, 1935; Onaka & Takahashi, 1968) indicates a band gap of 7·3 eV, which makes this impossible. Careful electrolysis experiments on ice (Decroly *et al.* 1957) confirm that the current is totally ionic, to within the accuracy of the measurement (about 5 per cent) and indicate protons as the ionic species involved. This still leaves open the question of whether ion states or orientational defects are the entities determining conductivity and indeed we shall see presently that the argument leading to a physical description of the conductivity process is not simple. However, the high-pressure measurements of Chan *et al.* (1965), already referred to, show that the static conductivity increases by about a factor 3 as the pressure is raised to 2000 bars, indicating a negative activation volume of $-11 \pm 3\ \text{cm}^3\ \text{mole}^{-1}$. As discussed before, this indicates quite unambiguously that it is the ion states which control the static conductivity. This fact was assumed, on the basis of rather less direct evidence, in some of the experimental work on point defects discussed in chapter 7.

9.4. Theoretical treatment

The currently accepted theory which describes these relaxation and conductivity phenomena is the result of contributions from many people, including Bjerrum (1951), Gränicher *et al.* (1957, 1963), Onsager & Dupuis (1960) and Jaccard (1959, 1964, 1965). The essence of the theory can be seen from a simple billiard-ball model of the molecular processes involved, while a detailed consideration of some of the mechanisms requires, as we shall see, very much greater sophistication. The treatment given in this section is essentially that of Jaccard (1959).

We recognize from the outset that there are four characteristic defects in the ice structure and treat them all in the theory. We have already discussed the formation and energetics of these defects in chapter 7; let us now see how they move and interact. Figure 9.5(*a*) shows the motion of a D-defect along a chain of

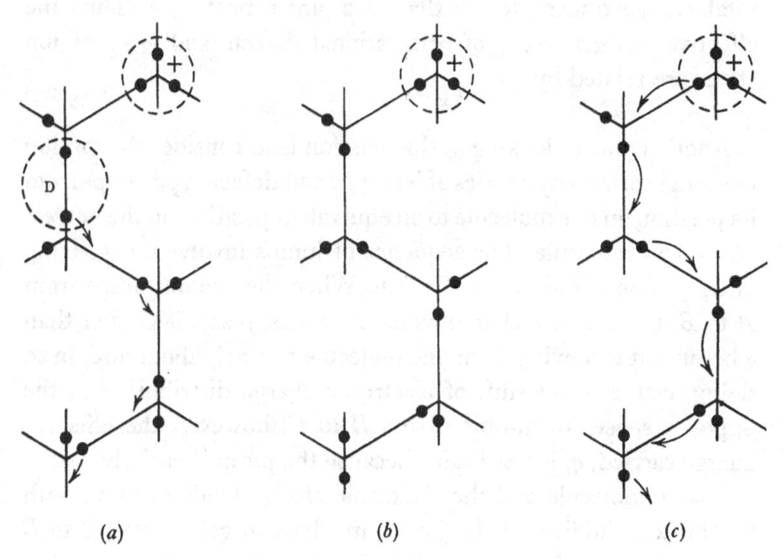

Fig. 9.5. Proton jumps associated with the motion of a D-defect along a chain of bonds, as in (*a*), leave this chain polarized, as in (*b*), so that no other D-defect can traverse the same bonds in the same direction. The proton jumps associated with motion of a positive ion state through the same chain of bonds, as in (*c*), restore the original proton configuration.

bonds which leads to the situation shown in (*b*), where all these bonds are polarized so that it is impossible for another D-defect to traverse them in the same direction or an L-defect in the opposite direction. These defects alone cannot, therefore, sustain a steady proton current. Figure 9.5(*c*), however, shows the effect produced by a positive ion state travelling along the same path followed by the D-defect. It is obvious that, by its passage, it resets all the bond orientations to their original states (*a*). The same effect would have been produced by a negative ion state moving in the opposite direction.

It is thus clear that neither orientational defects nor ion states can by themselves account for static electrical conductivity. Instead equal currents, measured in terms of the number of defects transported, must be carried by each mechanism so as to maintain the polarization of the ice structure at its equilibrium value. When both an orientational defect and an ion state move through the crystal in this way, leaving it in its original state, the total charge transported is that of a single proton, e. Thus the effective charges $\pm e_{DL}$ of orientational defects and $\pm e_{\pm}$ of ion states are related by

$$e_{\pm}+e_{DL} = e. \tag{9.23}$$

Another way of looking at this relation is to consider the motion of a single proton by a series of ion-type and defect-type jumps from its position on one molecule to an equivalent position on the nearest equivalent molecule. The sequence of jumps involved, neglecting other protons, is shown in fig. 9.6. When the proton jumps from A to B, the effective charge which it carries, p say, is greater than e because it is moving from one molecule to a neighbour and, in so doing, causes a net shift of electronic charge distribution in the opposite sense. In jumping from B to C, however, the effective charge carried, q, is less than e because the jump is entirely within the same molecule and the electronic charge tends to move with the proton. Adding all the jumps involved in going from A to E and noting that the total result is the transport of a single proton between these two crystallographically equivalent positions, we find

$$ap+2bq = er_0, \tag{9.24}$$

where the distances a, b and r_0 are as shown in fig. 9.6. The effective charges p and q are those which interact with the electric field and determine the mobility of the ion state or orientational defect, since the usual adiabatic approximation regards the electrons as always moving quickly enough to be in their equilibrium configuration. From (9.23) and (9.24) and consideration of the motions involved, it is clear that

$$|e_{\pm}| = pa/r_0, \quad |e_{DL}| = 2qb/r_0. \tag{9.25}$$

We can now write down general expressions for the current densities carried by the various defects in an electric field E, assuming the concentrations n_i and the temperature to be uniform. The flux of D-defects, for example, is given by

$$j_D = n_D \mu_D E + \Omega_D n_D \int_0^t (j_+ - j_- + j_L - j_D)\,dt, \qquad (9.26)$$

where μ_D is the D-defect mobility and Ω_D is a quantity measuring the extent to which the currents which have already passed through crystal may block the paths of the D-defects. Three similar

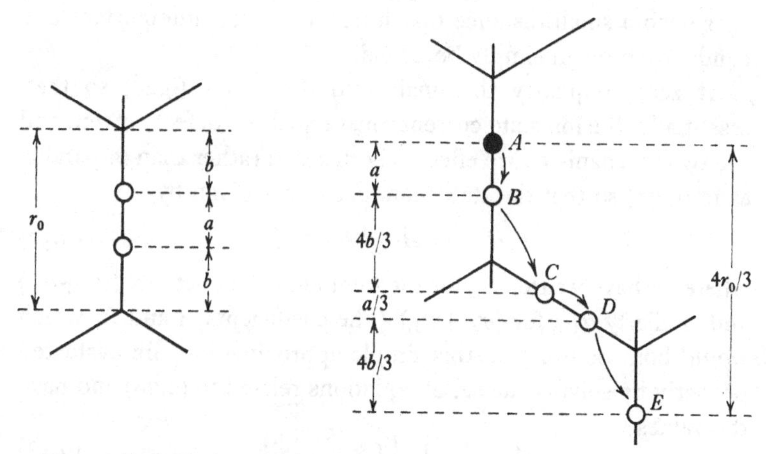

Fig. 9.6. Individual proton jumps between allowed positions on the bonds. The factors 1/3 and 4/3 arise from the tetrahedral geometry.

equations may be written for the current densities j_L, j_+ and j_-. The important quantities Ω_i and the mobilities μ_i and diffusion coefficients D_i have been evaluated by Jaccard (1959) in terms of the lattice geometry and occupation statistics and the equilibrium jump frequencies ν_i^0 for the various defects. The results are

$$\begin{aligned} \mu_{\pm} &= (1/6) r_0 \nu_{\pm}^0 ap/kT, & \mu_{DL} &= (2/9) r_0 \nu_{DL}^0 bq/kT, \\ D_{\pm} &= (1/6) r_0^2 \nu_{\pm}^0, & D_{DL} &= (1/9) r_0^2 \nu_{DL}^0, \\ \Omega_{\pm} &= (16/9\sqrt{3}) r_0^3 \nu_{\pm}^0, & \Omega_{DL} &= (1/\sqrt{3}) r_0^3 \nu_{DL}^0. \end{aligned} \qquad (9.27)$$

From equation (9.26) and the other three of its set it is now possible to derive the total electrical behaviour for, if E varies as $\exp(i\omega t)$, then the currents j_i vary similarly and (9.26) can be written

$$j_D = n_D \mu_D E + \Omega_D n_D (j_+ - j_- + j_L - j_D)/i\omega. \qquad (9.28)$$

At infinite frequency the second term vanishes for each defect and we have a simple set of equations of the form

$$j_i = n_i \mu_i E, \tag{9.29}$$

which lead to the high-frequency electrical conductivity

$$\sigma_\infty = \sigma_+ + \sigma_- + \sigma_D + \sigma_L, \tag{9.30}$$

where

$$\sigma_i = n_i \mu_i e_i. \tag{9.31}$$

This implies that, at such high frequencies, each carrier moves over such a small distance that interactions are unimportant and conductivities can simply be added.

At zero frequency the final term dominates (9.28) so that, essentially, the ion state current must equal the defect current and the two mechanisms are effectively in series rather than in parallel as in (9.30) so that the static conductivity is given by

$$\sigma_s^{-1} = A\sigma_\pm^{-1} + B\sigma_{DL}^{-1}, \tag{9.32}$$

where we have written $\sigma_\pm$ for the total ionic conductivity $(\sigma_+ + \sigma_-)$ and similarly σ_{DL} for $(\sigma_D + \sigma_L)$. The coefficients A and B, which would both be unity on this simple approximation, are evaluated properly by solving the set of equations related to (9.28) and have the values

$$A = \left(\frac{e_\pm}{e}\right)^2 \left(1 + \frac{10}{27}\frac{e_{DL}}{e}\right)^{-1}, \tag{9.33}$$

$$B = \frac{128}{27}\left(\frac{e_{DL}}{e}\right)^2 \left(1 + \frac{10}{27}\frac{e_{DL}}{e}\right)^{-1}.$$

From (9.32) the value of the static conductivity is determined by whichever of the two defect mechanisms is the less efficient. In pure ice this turns out to be that associated with ion states.

Solution of the set of equations (9.28) for finite frequency also gives an expression for the complex dielectric constant. With the subtraction of the static conductivity σ_s and addition of the dielectric component ϵ_∞ which is neglected in the present analysis, this has the observed Debye form (9.17). The static dielectric constant ϵ_s is given by

$$\epsilon_s - \epsilon_\infty = \frac{\left(\frac{\sigma_\pm}{e_\pm} - \frac{1}{2}\frac{\sigma_{DL}}{e_{DL}}\right)\left(\frac{32}{3\sqrt{3}}\frac{\sigma_\pm}{e_\pm} - \frac{3\sqrt{3}}{2}\frac{\sigma_{DL}}{e_{DL}}\right)}{r_0 kT\left(\frac{32}{3\sqrt{3}}\frac{\sigma_\pm}{e_\pm^2} + \frac{3\sqrt{3}}{4}\frac{\sigma_{DL}}{e_{DL}^2}\right)^2} \tag{9.34}$$

and the relaxation time τ by

$$\tau = \frac{1}{r_0 kT}\left(\frac{32}{3\sqrt{3}}\frac{\sigma_\pm}{e_\pm^2} + \frac{3\sqrt{3}}{4}\frac{\sigma_{DL}}{e_{DL}^2}\right)^{-1}. \tag{9.35}$$

It is worth emphasizing that, though the dielectric properties are treated here in terms of the motion of defects, the mechanism involved is essentially that of molecular reorientation so that the theory is really a kinetic version of the static treatment considered earlier in this chapter.

From Jaccard's discussion of jump probabilities in the ice structure, there is no apparent anisotropy in electrical behaviour except for that introduced by the almost negligible departure of the molecular environment from exact tetrahedral geometry. The observed small anisotropy in ϵ_s must therefore be attributed to an anisotropy in the effective charges $e_\pm$ and e_{DL}, which might well occur through their large electronic polarization component, or to an anisotropy in some of the higher-order terms which were omitted in determining the quantities in (9.27).

We have already discussed the high-pressure measurements of Chan *et al.* (1965) which show that the static conductivity σ_s is controlled by the behaviour of ion states while the dielectric relaxation time τ is controlled by orientational defects. Equations (9.32) and (9.35) therefore lead us to the conclusion that, for pure ice, $\sigma_\pm \ll \sigma_{DL}$. This is borne out by the work on ice crystals containing impurities which we shall discuss in the next section, and indeed that work led to this conclusion several years before the high-pressure results were available.

With the simplification allowed by this inequality in equations (9.30), (9.32), (9.34) and (9.35) it is now possible to determine a variety of important quantities. The temperature-dependence of the static conductivity gives an activation energy which is equal to the sum of half the energy $H_\pm^f$ for formation of a pair of ion states and $H_\pm^t$, the activation energy for transport of the more mobile ion state. Thus

$$\tfrac{1}{2}H_\pm^f + H_\pm^t = 0{\cdot}48 \pm 0{\cdot}07 \text{ eV}. \tag{9.36}$$

Similarly the activation energy associated with the dielectric relaxation time gives

$$\tfrac{1}{2}H_{DL}^f + H_{DL}^t = 0{\cdot}575 \pm 0{\cdot}03 \text{ eV}. \tag{9.37}$$

When the ion-state concentrations determined by Eigen *et al.* (1964) are combined with measured static conductivities, the mobility of the positive ion state is found to have the anomalously large value of 0·075 $cm^2 V^{-1} s^{-1}$. Such a high mobility can only be the result of some sort of quantum-mechanical tunnelling process, as we shall see later. For such a process $H^t \simeq 0$ so that, from (9.36), we conclude

$$H^f_{\pm} = 0{\cdot}96 \pm 0{\cdot}14 \text{ eV}. \tag{9·38}$$

Apart from this, the energies of formation and transportation cannot be further separated without the aid of data on doped crystals.

If the anisotropy of the static dielectric constant is attributed solely to anisotropy in the effective defect charges, then their values are (Jaccard, 1964)

$$\begin{aligned} e_{DL\parallel} &= 0{\cdot}55e, \quad e_{DL\perp} = 0{\cdot}51e, \\ e_{\pm\parallel} &= 0{\cdot}45e, \quad e_{\pm\perp} = 0{\cdot}49e. \end{aligned} \tag{9·39}$$

If the anisotropy is neglected, then appropriately weighted mean values may be used.

It is useful to note that the original equations (9.26) for the defect currents can be simply generalized to take account of thermal gradients and carrier concentration gradients in the crystal by adding, in the case of D-defects, a term $-\text{grad}(D_D n_D)$ to the right-hand side. We shall return to this when considering thermoelectric effects. The whole theory may, however, be reformulated with advantage using the methods of irreversible thermodynamics (Jaccard, 1964). The results obtained are closely similar to those set out above but can be more easily generalized to take account of other effects.

One such effect which has so far been omitted from our discussion is the electrostatic interaction between defects. This effect, which was considered in detail by Onsager & Dupuis (1960) and is also treated by Jaccard (1964) and by Dougherty (1965), becomes important when one type of carrier is in considerable excess and the motion of the minority carriers is being studied, for the majority carriers can then shield the other species electrostatically. Since these effects do not cause any very appreciable change in the final conclusions in the case of simple conductivity or dielectric measurements, we shall not consider them further here.

9.5. Ice crystals containing impurities

Although the treatment given in the previous section was for pure ice, there is nothing in the derivation of the results (9.30)–(9.35) restricting them to this case except for the carrier concentrations n_i, which are related to the partial conductivities σ_i by (9.31). Those results may therefore be applied to ice crystals containing known amounts of particular impurities simply by making the necessary adjustments to the concentrations n_i. Any impurities which leave the concentrations of ion states and valence defects unaltered will have negligible effect on the electrical properties.

From (9.32), which shows that σ_s is determined by the least effective conductivity mechanism, it is evident that this static conductivity may be quite critically dependent upon the presence of proton-donor or proton-acceptor impurities and, in this sense, the behaviour of ice is quite analogous to that of electronic semiconductors like germanium or silicon. Mole fractions of impurity which are significant are of the order of 10^{-9}. The high-frequency conductivity σ_∞ and relaxation time τ are, from (9.30) and (9.35), rather less sensitive to impurity content than is σ_s but are still directly affected at higher doping levels. The static dielectric constant ϵ_s depends on purity in a rather complex way, as we shall discuss presently.

The simplest experiments to interpret are those in which a proton acceptor such as hydrogen fluoride or a proton donor like ammonia is added to the crystal. Considering hydrogen fluoride first, we have already seen that, being very close in size to a water molecule, this impurity enters the ice structure substitutionally, bringing with it an L-defect. This defect may be liberated thermally and then moves freely through the lattice. Similarly there may be a thermal ionization process liberating an H_3O^+ ion state and leaving an immobile F^- ion.

Steinemann (1957) has measured the electrical properties of ice doped with hydrogen fluoride and finds that at $-10\,^\circ$C the static conductivity σ_s is proportional to the square root of hydrogen fluoride concentration

$$\sigma_s(\mathrm{HF}) = 1{\cdot}6 \times 10^{-15}(n_{\mathrm{HF}})^{\frac{1}{2}}\ \Omega^{-1}\ \mathrm{cm}^{-1} \qquad (9.40)$$

for n_{HF} in the range 10^{15}–10^{19} cm^{-3}. For lower concentrations σ_s tends towards independence of n_{HF}. Since ion states control the static conductivity, this implies that $n_+ \propto (n_{HF})^{\frac{1}{2}}$, which is just what would be expected for a slightly ionized impurity, for which the law of mass action gives

$$n_+ n_{F-} = n_+^2 = k_{HF} n_{HF} \tag{9.41}$$

and, since $n_+ \ll n_{HF}$, n_{HF} is closely equal to the total added hydrogen fluoride concentration.

Jaccard (1959) has measured the conductivity of a heavily hydrogen-fluoride-doped crystal ($n_{HF} = 1{\cdot}7 \times 10^{18}$ cm^{-3}) over a wide temperature range and finds an activation energy of

$$0{\cdot}325 \pm 0{\cdot}005 \text{ eV}.$$

Since, as discussed before, the effective activation energy for proton transport is zero, we conclude that this is the energy for liberation of a positive ion state from a substitutional hydrogen fluoride molecule in ice.

Since $\sigma_\pm \ll \sigma_{DL}$ for pure ice and since n_+ increases rather slowly as n_{HF} is increased, we may expect from (9.30) that the high-frequency conductivity σ_∞ may be dominated by the effects of changes in σ_{DL}. Measurements by Jaccard (1959) show that, at -55 °C,

$$\sigma_\infty(\text{HF}) = 1{\cdot}6 \times 10^{-24} n_{HF} \ \Omega^{-1} \text{cm}^{-1} \tag{9.42}$$

for n_{HF} in the range 10^{17}–10^{18} cm^{-3}. Such a linear relation, which implies similar behaviour for n_L, can only occur if the binding energy of L-defects to hydrogen fluoride molecules is so small that essentially all defects are liberated. This binding energy is thus of order 10^{-2} eV or less and the measured activation energy for σ_∞, $0{\cdot}235 \pm 0{\cdot}01$ eV, must describe the mobility of L-defects. Using (9.37) we thus have

$$H_{DL}^f = 0{\cdot}68 \pm 0{\cdot}04 \text{ eV}, \quad H_L^t = 0{\cdot}235 \pm 0{\cdot}01 \text{ eV}. \tag{9.43}$$

The behaviour of the static dielectric constant as a function of hydrogen fluoride impurity concentration is complicated, as shown in fig. 9.7 from the work of Gränicher *et al.* (1957) and Steinemann (1957). The additional complexities caused by space-charge effects at the electrodes have been omitted. As n_{HF} is increased

ϵ_s falls from the value near 100 characteristic of pure ice to essentially its high-frequency value $\epsilon_\infty = 3{\cdot}2$. It then rises to about 25, falls again to about 3 and finally rises again at the highest hydrogen fluoride concentration at which homogeneous crystals can be made.

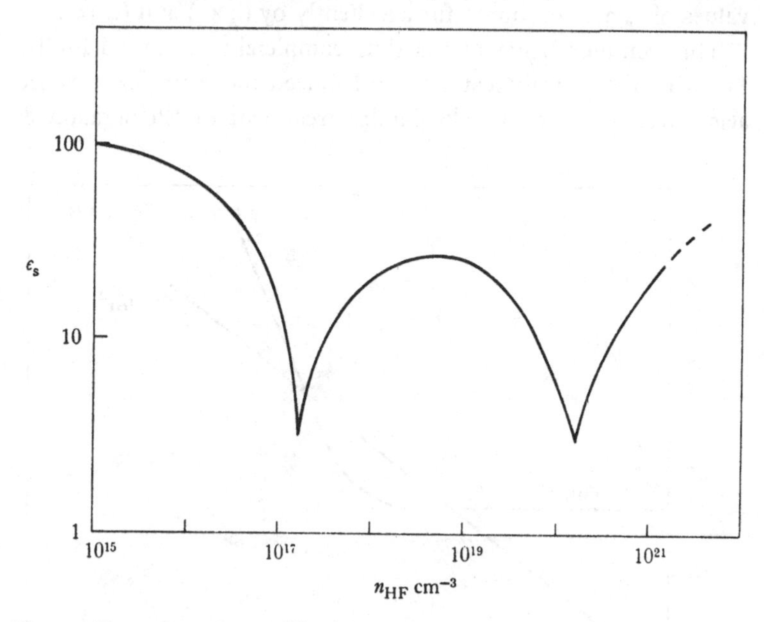

Fig. 9.7. Dependence, at −3 °C, of the static dielectric constant of ice upon the concentration n_{HF} of hydrofluoric acid impurity (after Steinemann, 1957).

This behaviour is simply accounted for in terms of the expression (9.34) for ϵ_s. If $\sigma_{DL} \gg \sigma_\pm$, as is the case for pure ice, then

$$\epsilon_s(\mathrm{DL}) \simeq \epsilon_\infty + \frac{4}{3\sqrt{3}} \frac{e_{DL}^2}{kTr_0} \simeq 100, \tag{9.44}$$

while, if $\sigma_\pm \gg \sigma_{DL}$,

$$\epsilon_s(\pm) \simeq \epsilon_\infty + \frac{3\sqrt{3}}{32} \frac{e_\pm^2}{kTr_0} \simeq 21, \tag{9.45}$$

and if $\sigma_\pm \simeq \frac{1}{2}\sigma_{DL}$ both brackets in the numerator are close to vanishing and $\epsilon_s \simeq \epsilon_\infty$. This cancellation of dielectric effects can also be seen in principle from the early equation (9.28), from which it is apparent that the imaginary term in the conductivity vanishes when $j_\pm$ and j_{DL} are equal. The variations in $\sigma_\pm$ and σ_{DL} are described by (9.40) and (9.42) when these have been corrected by adding the partial conductivities characteristic of pure ice. Correcting to -3 °C, the situation is then as shown in fig. 9.8. The actual positions of the transition points calculated from (9.40) and (9.42) are only in rough agreement with experiment but the absolute values of ϵ_s are accounted for excellently by (9.44) and (9.45).

This complex behaviour is thus completely accounted for by Jaccard's theoretical treatment and indeed the main features are also covered by the much simpler treatment of Steinemann &

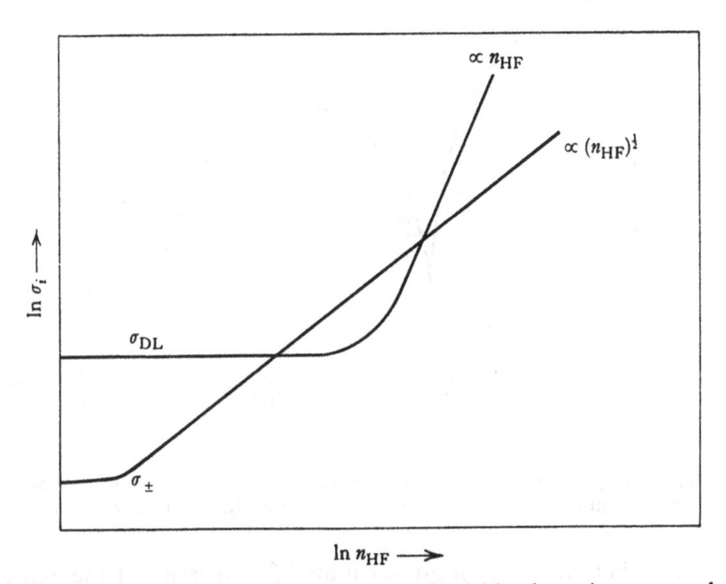

Fig. 9.8. Schematic behaviour of the partial conductivities due to ion states and to orientational defects in an ice crystal containing hydrogen fluoride.

Gränicher (1957), which considered only the majority carriers in each region. It is somewhat more difficult to picture the polarization processes in terms of molecular orientations. In the regions where orientational defects provide the relaxation mechanism, their motion in the applied field simply reorients molecules or, if the effective charge associated with the defects is included, then this contributes a polarization in the same direction. In regions where ion states provide the relaxation mechanism, however, motion of these states in the direction of the field tends to orient molecular dipoles antiparallel to the field. The polarizations produced by molecular orientation and ion-state motion are thus in opposition and the effective dielectric constant is greatly reduced. This was realized by Bjerrum (1951) and was one of the reasons leading to his postulation of the existence of L- and D-defects.

When we come to consider ice crystals doped with ammonia we might expect the experimental situation to be very similar to that discussed above, with negative ion states replacing positive and D-defects replacing L-defects. This is broadly true, though there are some interesting differences. The experimental work, which has not been as extensive as for hydrogen fluoride ice, is due to Iribarne *et al.* (1961), Levi & Lubart (1961) and Levi *et al.* (1963).

The static conductivity increases as $(n_{NH_3})^{\frac{1}{2}}$ for n_{NH_3} less than about 2×10^{18} cm^{-3}; for greater concentrations, σ_s is independent of ammonia concentration. The activation energy associated with σ_s is reported (Levi & Lubart, 1961) as 0·7 eV for $n_{NH_3} < 10^{16}$ cm^{-3}, a value close to that for pure ice. For n_{NH_3} between 10^{16} and 10^{18} cm^{-3} the activation energy is not well defined, whilst above this concentration, where σ_s is independent of n_{NH_3}, it has the value $0{\cdot}37 \pm 0{\cdot}01$ eV. These results can be interpreted as implying that the energy for liberation of a positive ion state from ammonia in ice is about 0·7 eV while a D-defect is bound with an energy of about 0·3 eV. If this is the case then n_D will be very little affected by n_{NH_3} and, for n_{NH_3} greater than about 10^{18} cm^{-3}, $\sigma_\pm$ will cease to be the minority mechanism. σ_{DL} will then determine the static conductivity which will be insensitive to n_{NH_3}. If this is a correct interpretation then ϵ_s should go through a minimum near

$$n_{NH_3} \simeq 10^{18} \text{ cm}^{-3}$$

and then rise to a value near 21, but there should not be any second minimum as with hydrogen fluoride. The experiments suggest that this happens but are not extensive enough to prove the point.

The behaviour of much more complex impurity systems can be treated in the same sort of way provided the relevant dissociation constants for production of ion states and orientational defects are known. The case of ammonium fluoride is of particular interest since we have already discussed its two components in detail. If the dissociation properties of ammonia and hydrogen fluoride were identical, then addition of ammonium fluoride to ice would produce no first-order effects. However hydrogen fluoride produces both ion states and orientational defects more easily than does ammonia, so that the net effect of ammonium fluoride is to add positive ion states and L-defects. This might be expected to be reflected in the behaviour of the dielectric relaxation time τ, but, while some workers report that τ varies approximately inversely with the concentration of ammonium fluoride (Zaromb & Brill 1956), others (Dengel *et al.* 1966) have found no such dependence.

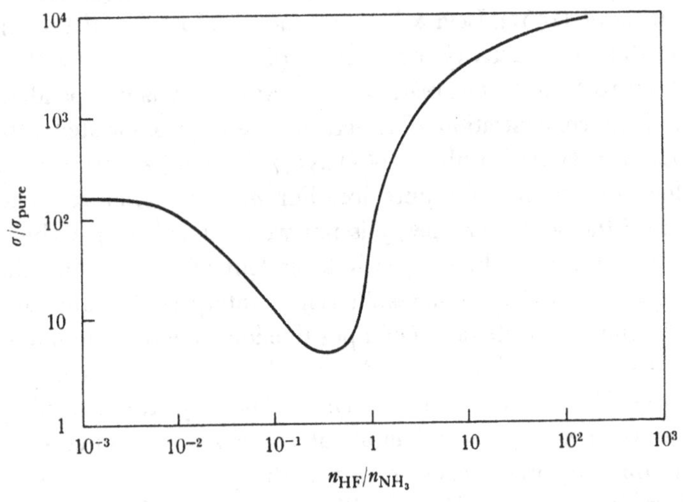

Fig. 9.9. Static electrical conductivity of an ice crystal as a function of the ratio of concentrations of hydrogen fluoride and ammonia (after Levi *et al.* 1963). The experimental uncertainty is about a factor of 3.

The two components of ammonium fluoride behave in very much the same way as do donor and acceptor impurities in electronic semiconductors, leading to more or less complete compensation. This behaviour is illustrated, as far as static conductivity is concerned, by the careful study of Levi *et al.* (1963) in which the ratio of hydrogen fluoride to ammonia concentrations was varied over a wide range. The experimental results are shown in fig. 9.9 and illustrate the expected behaviour. This can be calculated by the normal chemical methods used for a solution containing a weak acid and a weak base but, to obtain reasonable agreement with experiment, it was found necessary to assume that the positive ion state mobility is $\mu_+ \simeq 2 \times 10^{-2}$ cm^2 V^{-1} s^{-1} and that the mobility ratio μ_+/μ_- lies in the range 1–5. These values are rather different from the more generally accepted ones of Eigen *et al.* (1964) discussed in chapter 7 but the disagreement is not extreme.

Investigations have been made of the behaviour of various other impurities in ice by Iribarne *et al.* (1961) and by Levi & Arias (1964) but the details need not concern us here. For convenience the experimental data on hydrogen fluoride and ammonia are summarized in table 7.4 on page 159. From the deviations shown it is clear that many of these quantities have not yet been accurately determined.

9.6. Thermoelectric effects

The basic equations (9.26) for the motion of charge carriers in ice can easily be generalized, as we remarked before, to include the effects of temperature or concentration gradients, in which case they take the form

$$j_D = n_D \mu_D E - \frac{d}{dx}(D_D n_D) + \Omega_D n_D \int_0^\infty (j_+ - j_- + j_L - j_D)\, dt \qquad (9.46)$$

together with similar equations for the other carriers. This set of equations has been solved by Jaccard (1963) for the case of a uniform temperature gradient dT/dx giving rise in an isolated crystal to a potential gradient dV_T/dx. The homogeneous thermoelectric power, or thermopower, Q_H, is given by

$$Q_H = -\frac{dV_T}{dT} = \frac{\Gamma}{eT}\left(27\,\frac{n_+(H_+^f + H_+^t) - \phi^\pm n_-(H_-^f + H_-^t)}{(n_+ + \phi^\pm n_-)} + 32\,\frac{n_D(H_D^f + H_D^t) - \phi^{DL} n_L(H_L^f + H_L^t)}{(n_D + \phi^{DL} n_L)}\right), \qquad (9.47)$$

where H_i^f and H_i^t are the activation energies associated with formation and transport of the various species, $\phi^\pm$ and ϕ^{DL} are the ratios

$$\phi^\pm = D_-/D_+, \quad \phi^{DL} = D_L/D_D = \mu_L/\mu_D, \qquad (9.48)$$

and the factor Γ is given by

$$1/\Gamma = 27\left(\frac{e_\pm}{e}\right)\frac{n_+ + n_- \mu_-/\mu_+}{n_+ + n_- D_-/D_+} + 32\left(\frac{e_{DL}}{e}\right). \qquad (9.49)$$

The Einstein relation between diffusion coefficient and mobility is assumed to apply for the motion of orientational defects but may not apply for the quantum tunnelling motion of ion states. (It should be noted in passing that some workers define Q_H with a sign opposite to that used here.)

The physical origin of this thermoelectric effect is clear. If one end of a crystal is heated then the concentrations and diffusion coefficients of carriers in that region will increase, the relative increases being determined by the activation energies H_i^f and H_i^t respectively. Each carrier will tend to move down the concentration gradient toward the cold end of the crystal, until the charge which it carries builds up a great enough electrostatic field to cause a mobility flow which exactly balances the diffusion flow. In an ice crystal in isolation, not only must the total electrical current balance to zero but also the currents due to ion states and orientational defects separately, so that the crystal polarization can remain constant. Since Γ, defined by (9.49), varies little with carrier concentration, the homogeneous thermopower Q_H consists of the sum of two essentially independent terms referring to ion states and orientational defects separately. The contribution of each of these terms depends on the ratio, n_+/n_- or n_D/n_L, of the carrier concentrations involved and not upon the absolute values of the carrier concentrations.

The first experimental determination of the homogeneous thermopower of nominally pure ice was made by Latham & Mason (1961*a*), who found a value of 2 mV degC^{-1}, the positive sign indicating that the warmer end of the ice crystal became negatively charged. The thermopower was apparently not very sensitive to addition of impurities such as hydrogen fluoride or carbon dioxide. Above -7 °C the measured value of Q_H became greater than this and depended as well on the surface-to-volume ratio of the ice specimens used, implying that some sort of surface effect was involved (Latham, 1963).

The important thing about this measurement, which antedated Jaccard's theoretical discussion, is the large magnitude of the observed thermopower, a magnitude comparable to that for semiconductors and very much greater than the tens of microvolts per degree typical of metals. This circumstance led Latham & Mason

(1961 *b*) to base a theory of the electrification of thunderclouds upon thermoelectric charging of ice splinters ejected from cloud droplets freezing on collision with soft hail, a theory which seems able to explain many, though perhaps not all, of the observed phenomena.

Another consequence of the large value of Q_H is that the effect of temperature gradients in metallic leads to the ice crystal can be virtually neglected, though the use of metallic electrodes itself introduces problems because of the unknown temperature variation of any interfacial potentials at the electrodes. These extraneous effects seem to be negligible for carefully prepared electrodes, as is confirmed by the experiments of Latham (1964) and of Brownscombe & Mason (1966), who suspended an ice crystal in a temperature gradient and then measured the potential gradient by an inductive method. The result which Brownscombe and Mason found, $Q_H = 2{\cdot}3 \pm 0{\cdot}3$ mV degC^{-1} at -20 °C for nominally pure ice crystals, is in good agreement with the earlier value, while Latham found a similar value with a wider uncertainty.

The first systematic study of the effect of proton donor or acceptor impurities on the thermopower was made by Bryant & Fletcher (1965) using either hydrogen fluoride or ammonia as impurity and with metallic electrodes. Unfortunately their ice specimens were polycrystalline and, because there is inevitably a great deal of impurity segregation at grain boundaries, the true concentrations within crystallites were probably one to two orders of magnitude less than those shown. Bryant (1967) later repeated these measurements with carefully grown single crystals and found behaviour of the same general pattern but differing somewhat in the magnitudes involved. These results, together with those on nominally pure ice, are summarized in fig. 9.10. Also shown are measurements by Takahashi (1966) on single crystals from the Mendenhall Glacier, which showed some anisotropy with $Q_{\parallel} < Q_{\perp}$, and on two polycrystalline samples containing hydrogen fluoride. All workers reported anomalous behaviour, sometimes even a change in the sign of Q_H, above about -12 °C.

From fig. 9.10 and taking into account the way in which the experimental data for polycrystalline samples should be adjusted to allow for segregation effects, it can be seen that there is a large measure of agreement between the different experiments. This behaviour is very much what would be predicted from the form of (9.47) and can be understood as follows.

The contributions of ion states and orientational defects to Q_H are very nearly independent. Since the intrinsic concentration of L- and D-defects is $\sim 10^{15}$ cm^{-3}, their concentrations will not be affected for impurity concentrations less than $\sim 10^{15}$ cm^{-3}, so all

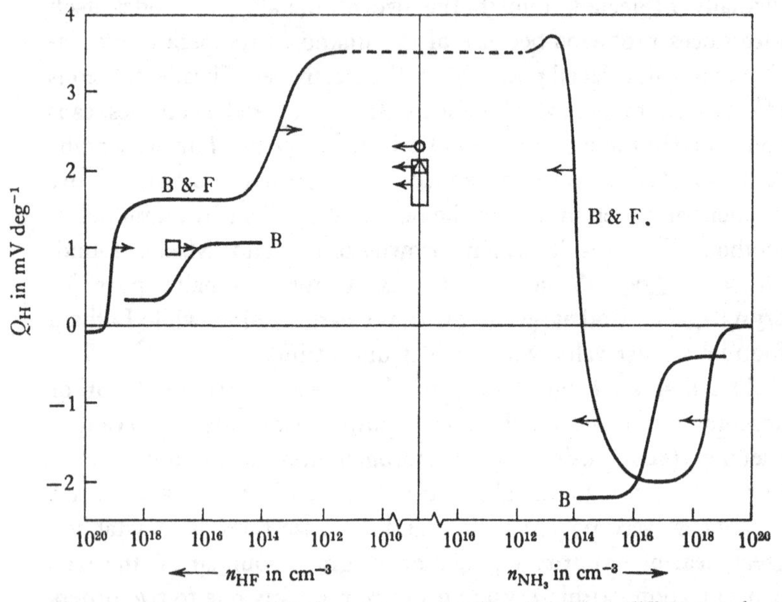

Fig. 9.10. The homogeneous thermopower Q_H of ice according to different workers. Arrows show the direction in which individual experimental results should be adjusted because of the conditions of the experiment. △, Latham & Mason (1961*a*); B & F, Bryant & Fletcher (1965); ○, Brownscombe & Mason (1966); □, Takahashi (1966); B, Bryant (1967).

variations in Q_H for smaller concentrations must be caused by ion states, whose intrinsic concentration is only $\sim 10^{10}$ cm^{-3}. For pure ice, since $\phi^{\pm} = D_{-}/D_{+} \ll 1$ and the transport energies $H^{t}_{\pm}$ are zero, the ionic contribution to Q_H is approximately $27\Gamma H^{f}_{+}/eT$, which is positive and determined in magnitude by H^{f}_{+}, the energy of formation of an H_3O^+ state by dissociation of a water molecule. As hydrogen fluoride is added positive ion states remain in the majority but H^{f}_{+} becomes the smaller energy required to dissociate a hydrogen fluoride molecule, so Q_H falls, though the ionic contribution is still positive.

If, on the other hand, ammonia is added, then, for ammonia concentration greater than about 10^{10} cm^{-3}, n_- begins to increase while H^f_- falls to the value appropriate for dissociation of ammonia. Depending upon the change in H^f_-, the product $n_- H^f_-$ may initially fall slightly, giving a small increase in Q_H, but then rises linearly. This term dominates the positive ion contribution at a concentration determined by $\phi^{\pm}$, and the total ion-state contribution falls to zero and then saturates at the negative value $-27\Gamma H^f_-/eT$.

For concentrations of ammonia or hydrogen fluoride less than about 10^{15} cm^{-3} the contribution made to Q_H by orientational defects is constant and has a sign determined primarily by the mobility ratio $\phi^{DL} = D_L/D_D$ which is not well known from other work. When extreme concentrations of hydrogen fluoride are added, L-defects dominate and the contribution of orientational defects to Q_H falls from the pure ice value to approximately

$$-32\Gamma(H^f_L + H^t_L)/eT.$$

Similarly, for extreme ammonia concentrations, the contribution rises from the pure ice value to

$$32\Gamma(H^f_D + H^t_D)/eT.$$

When the various dissociation energies for hydrogen fluoride and ammonia, as found from other experiments, are substituted into (9.47) with ϕ^{DL} as an unknown parameter, then good general agreement in shape is obtained if

$$\phi^{DL} \equiv D_L/D_D = 1 \cdot 5 \pm 0 \cdot 2. \tag{9.50}$$

The lack of reliable measurements in the very low impurity concentration regions means that an accurate value for $\phi^{\pm}$ cannot be found, but the measured value for nearly pure ice indicates

$$\phi^{\pm} \equiv D_-/D_+ < 0 \cdot 1, \tag{9.51}$$

which is in agreement with other estimates.

The absolute magnitude of the calculated value of Q_H depends on the way in which the relation between diffusion coefficient and mobility for ion states differs from the classical Einstein form. If we introduce parameters θ_+ and θ_- and write

$$\mu_{\pm} = \frac{e_{\pm}}{kT} D_{\pm} \theta_{\pm} \tag{9.52}$$

then, from (9.47) and (9.49), Q_H is affected through Γ only. Comparison with the experimental results (Bryant, 1967) suggests that

$$\theta_+ \simeq \theta_- \simeq 0.5 \pm 0.2. \tag{9.53}$$

This deviation of θ_+ and θ_- from unity implies some sort of correlation in successive proton jumps, which was neglected in the derivation leading to equations (9.27). We shall return to discuss this presently.

To conclude this discussion of thermoelectric effects in ice, we should briefly relate the homogeneous thermopower Q_H to some of the thermoelectric coefficients more common in electronic conductors. The whole subject can be treated by the methods of irreversible thermodynamics (Jaccard, 1964; Ziman, 1960, pp. 270–5) but we shall follow instead the more transparent microscropic treatment (Jaccard, 1963).

In a thermoelectric circuit such as that used for measuring Q_H, the current in the ice is carried by protons and that in the metallic leads by electrons and we noted that complications might arise at the electrodes. A much simpler circuit for analysis consists of two ice elements containing different impurities connected in series and with their junctions maintained at different temperatures, as shown in fig. 9.11.

To separate out the effect of the impurity changes at the junctions, we consider our basic equations (9.46) for the case where the temperature is constant but the carrier concentrations vary. These equations can then be integrated directly, assuming no net current to flow, to give the electrochemical potential V_C as

$$V_C = -\frac{kT}{e}\frac{\Gamma}{2}[27\ln(n_+/n_-)+32\ln(n_D/n_L)], \tag{9.54}$$

where V_C is taken as zero for pure ice. This relation has been checked experimentally by Brownscombe & Mason (1966) and reasonably good agreement with theory is obtained when Γ is calculated with $\theta_\pm \simeq 0.5$ as before.

The difference in electrochemical potential between two differently doped ice crystals is clearly temperature-dependent, since the ratios (n_+/n_-) and (n_D/n_L) depend strongly upon temperature, and these junction potentials must be taken into account in calculating the open-circuit thermocouple potential. If we define the total or inhomogeneous thermopower by

$$Q = Q_{\mathrm{H}} + \frac{dV_{\mathrm{C}}}{dT} \tag{9.55}$$

then the total thermocouple potential V, as measured in the Seebeck effect for ice specimens A and B with junction temperatures T_1 and T_2, is

$$V = \int_{T_1}^{T_2} (Q_B - Q_A)\,dT. \tag{9.56}$$

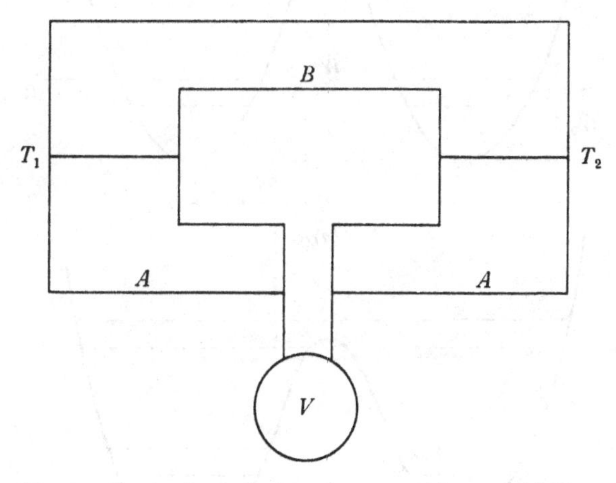

Fig. 9.11. Thermoelectric circuit for ice: thermocouple potential V produced by the Seebeck effect for ice specimens A and B of different impurity content, with junction temperatures T_1 and T_2.

The Peltier coefficient Π can be found from Kelvin's relation to be

$$\Pi = -QT \tag{9.57}$$

while the Thomson coefficient σ is

$$\sigma = T dQ/dT. \tag{9.58}$$

9.7. Theory of proton mobility

From the discussion of the last two sections it is clear that the microscopic processes governing the motion of protons from one defect site to another are of prime importance in determining the macroscopic electrical behaviour of an ice crystal. We have stated that the proton jumps involved in the migration of orientational defects are essentially classical while those for the motion of ion states occur by a quantum tunnelling process. Let us now examine these assertions more carefully.

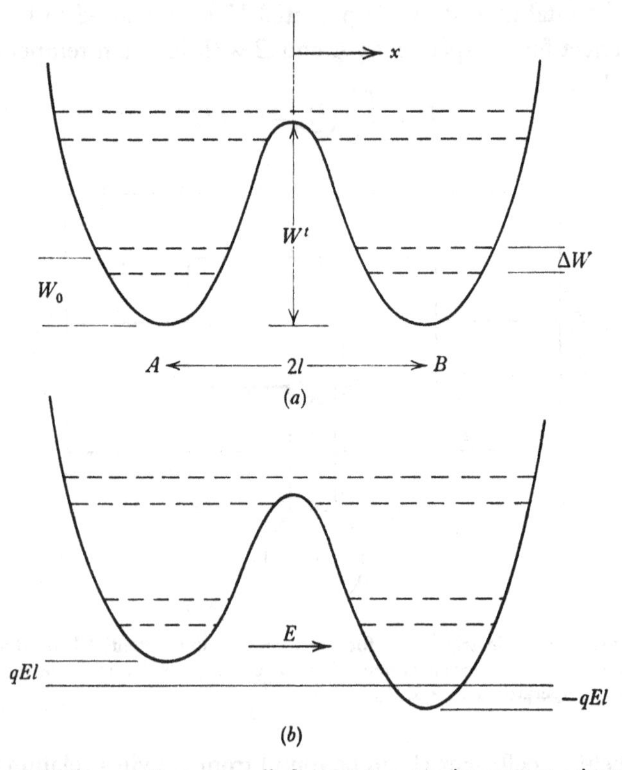

Fig. 9.12. (*a*) Symmetric energy wells for a proton at an ion-state or orientational defect. (*b*) Effect upon these wells of the application of an electric field E in the x-direction as shown. Energy levels are shown in appropriate positions for a quantum-mechanical calculation.

For any of the proton jump processes in which we are interested, the problem can initially be simplified to the motion of a single proton in a double-well potential of the kind shown in fig. 9.12(*a*). In the absence of an external field, the two wells are symmetrical and the proton is equally likely to be found in either. When a field is applied, however, as in fig. 9.12(*b*), the potential is no longer symmetrical and there is a tendency for the proton to migrate in the direction of the field. Whether the proton behaviour is governed by classical or quantum considerations depends primarily upon the shape and height of the energy barrier between the wells. If these are such that there are many energy levels for the proton below the top of the barrier, then the behaviour will be classical, while, if the number of such levels is very small, quantum effects will predominate.

In our discussion we shall first look at the behaviour of this simple system in isolation and then see what changes are necessary when it is recognized to be part of a large crystal involving many identical units. The treatment cannot be definitive, since no comprehensive solutions have yet been derived, but most of the factors involved can be reasonably well appreciated.

Now consider the potential shown in fig. 9.12(*a*). Each well is roughly parabolic at low energies so that the quantum levels are $W_n \simeq (n+\frac{1}{2})h\nu$, where ν is the classical oscillation frequency. Each level is slightly split by the interaction between the two potentials, as we shall see later, and there is some change in ν when W_n is greater than the barrier height W^t. If $h\nu \ll W^t$ then the behaviour is manifestly classical, the relative probability of a proton in one of the wells having enough energy to surmount the barrier is $\exp(-W^t/kT)$ and the total jump rate in either direction is

$$P_{A\leftrightarrow B} = \frac{\nu \sum\limits_{W_n>W^t} \exp\left(\dfrac{-W_n}{kT}\right)}{\sum\limits_{W_n>0} \exp\left(\dfrac{-W_n}{kT}\right)} \simeq \nu \exp\left(\frac{-W^t}{kT}\right), \qquad (9.59)$$

where the final approximation comes from the assumption that ν does not change much with energy so that the energy levels are nearly equally spaced.

Now suppose we apply an electric field E along the line joining the wells so that the situation becomes as in fig. 9.12(*b*). If the distance between the wells is $2l$ and the effective charge on the proton, allowing for adiabatic electronic motion, is q, then the level shifts in the two wells are $\pm qlE$. The jump rates with and against the field then become

$$P_{A\to B} = \nu \exp[-(W^t - qlE)/kT],$$
$$P_{B\to A} = \nu \exp[-(W^t + qlE)/kT), \qquad (9.60)$$

so that the net jump rate in the direction of the field, assuming $qlE \ll kT$, is

$$P_E = P_{A \to B} - P_{B \to A} = \frac{\nu qlE}{kT} \exp(-W^t/kT). \qquad (9.61)$$

Now in an ice crystal each proton jump over a distance $2l$ as in fig. 9.12 effectively moves the defect concerned through some larger distance, say r. From fig. 9.6 these two distances can be related to characteristic spacings in the lattice for either the direct jumps associated with ion states or the oblique jumps characteristic of valence defects. More than this, however, not all jump directions make the same angle with the field, though in the unpolarized ice structure there is always a possible jump with a component in the field direction. This average is simply performed and leads to an average defect displacement parallel to the field which we can write as $\langle r \rangle$. From (9.61) the classical mobility is then

$$\mu_C = \frac{\nu ql\langle r \rangle}{kT} \exp(-W^t/kT). \qquad (9.62)$$

This is the sort of behaviour found experimentally for D- and L-defects but not for ion states.

Suppose, however, that the barrier between the two wells is very narrow and not very high so that there are only a few proton energy levels less than W^t; then it is no longer possible to treat the energy states in the two wells as independent and quantum coupling effects equivalent to tunnelling through the barrier must be considered. This problem has been discussed by Conway *et al.* (1956) but their treatment neglects quantization of energy levels in the wells and therefore cannot give quantitatively correct results in the present case. A more suitable treatment is that of Jaccard (1959), and our discussion is a somewhat modified version of this. We shall return to a more sophisticated discussion later.

The ground-state wave function for a proton in the double well shown in fig. 9.12(a) can be approximately represented by a linear combination of two ground-state harmonic oscillator wave functions ψ_A and ψ_B centred about the points A and B respectively. Thus

$$\psi = \psi_A + \lambda \psi_B. \qquad (9.63)$$

The value of λ can be determined by requiring that the expectation value of the energy be stationary and we find, as indeed is required by symmetry, that $\lambda = \pm 1$. The two normalized 'molecular orbitals' for the problem are thus

$$\psi_{\pm} = 2^{-\frac{1}{2}}(\psi_A \pm \psi_B) \tag{9.64}$$

and the energy level is split into a doublet, the lower state corresponding to the symmetric wave function ψ_+. The energies involved can be evaluated by standard methods (Coulson, 1961, chapter 4) and, to a sufficient approximation for our present purpose,

$$W_{\pm} = W_0 \pm \tfrac{1}{2}\Delta W, \tag{9.65}$$

where the splitting ΔW is equal to twice the matrix element of the Hamiltonian H between the two localized states

$$\Delta W = 2\langle\psi_A|H|\psi_B\rangle. \tag{9.66}$$

The same sort of combination and splitting occurs for higher energy levels.

The wave functions $\psi_{\pm}$ have time dependence determined by their energies $W_{\pm}$ so that, if we construct a wave packet

$$\psi = 2^{-\frac{1}{2}}(\psi_+ + \psi_-)$$

which, at $t = 0$, represents a particle in well A, its subsequent behaviour is given by

$$\begin{aligned}\psi(t) &= 2^{-\frac{1}{2}}[\psi_+ \exp(-iW_+t/\hbar) + \psi_- \exp(-iW_-t/\hbar)] \\ &= [\psi_A \cos(\Delta Wt/2\hbar) + \psi_B \sin(\Delta Wt/2\hbar)]\exp(-iW_0t/\hbar),\end{aligned} \tag{9.67}$$

so that the rate of proton jumps between the two wells when the proton energy is close to W_0 is

$$P_{A\leftrightarrow B} = \Delta W/h. \tag{9.68}$$

It is, of course, not possible to include a statistical weighting function $\exp(-\Delta W_n/kT)$ directly in the proton wave function, since this must actually be inserted in the density matrix. The total jump rate, however, amounts essentially to a summation of quantum contributions like (9.68) from energy levels below the barrier top together with a classical contribution like (9.59) from higher levels.

If now we apply an electric field E and suppose that the effective proton charge is q, then we have the situation shown in fig. 9.12(b). From simple perturbation theory, the positions and therefore the separations of the energy levels are not altered in first order, since the matrix elements $\langle\psi_\pm|qEx|\psi_\pm\rangle$ are zero. There will, however, be first-order corrections to the wave functions so that, neglecting higher states, $\psi_\pm \to \psi'_\pm$, where

$$\psi'_\pm = \psi_\pm \mp (qE\beta/\Delta W)\psi_\mp, \tag{9.69}$$

where β is the positive quantity

$$\beta = \langle\psi_+|x|\psi_-\rangle, \tag{9.70}$$

which, in the present approximation, is simply equal to l.

This change in the wave functions can be thought of classically as a difference in distribution of protons between the two wells. The mean proton position shifts from $\langle x\rangle = 0$, characteristic of the unperturbed wave functions, to

$$\langle x\rangle_\pm = \langle\psi'_\pm|x|\psi'_\pm\rangle = \mp 2qE\beta^2/\Delta W \tag{9.71}$$

for the states ψ_+ and ψ_- respectively. If the fractional populations of the two wells in these two states are $n_{A\pm}$ and $n_{B\pm}$ respectively, then the co-ordinate shift is equivalent to a population shift:

$$n_{A\pm}/n_{B\pm} = (l - \langle x\rangle_\pm)/(l + \langle x\rangle_\pm). \tag{9.72}$$

From (9.71), the centre of mass for the ψ_+ state shifts to the right while that for the ψ_- state shifts to the left. At a finite temperature, however, the ψ_+ state has a larger population than the ψ_- state, so the net shift is to the right towards well B. Combining the two states we therefore have for the fractional population in well A at temperature T

$$\frac{n_A}{n_A + n_B} = \frac{\cosh(\Delta W/2kT) - (2qE\beta^2/l\Delta W)\sinh(\Delta W/2kT)}{2\cosh(\Delta W/2kT)}. \tag{9.73}$$

Since this shift is an equilibrium one, it must be maintained by a modification of the jump rates in the directions $A \to B$ and $B \to A$ so that, at temperature T, the net jump rate in the direction of the field for these two levels becomes

$$P_E^0 = \frac{4q\beta^2 E}{hl}\tanh(\Delta W/2kT). \tag{9.74}$$

We can write down a net quantum tunnelling rate like this for each pair of levels with energy W_n less than W^t and these must be summed with a weighting $\exp(-W_n/kT)$ to allow for their relative populations at temperature T. In addition, levels above the barrier for which $W_n > W^t$ make essentially a classical contribution of appropriate weight to the total jump rate, so that the final expression is

$$P_E = \frac{\displaystyle\sum_{W_n<W^t} \frac{4q\beta_n^2 E}{hl} \tanh\left(\frac{\Delta W_n}{2kT}\right) \exp\left(-\frac{W_n}{kT}\right) + \sum_{W_n>W^t} \frac{\nu q l E}{kT} \exp\left(-\frac{W_n}{kT}\right)}{\displaystyle\sum_{W_n>0} \exp\left(-\frac{W_n}{kT}\right)}, \tag{9.75}$$

which is identical with the result obtained by Jaccard (1959).

Since the mobility is proportional to P_E/E, it is interesting to see how this quantity varies with the other parameters involved. Of particular interest is the variation of mobility with temperature and the transition between quantum and classical behaviour. In the low-temperature region for which $kT \ll \Delta W \ll W^t$, (9.75) simplifies to

$$P_E^0 \simeq 4q\beta^2 E/hl, \tag{9.76}$$

which gives a temperature-independent quantum-mechanically determined mobility. At a rather higher temperature such that $\Delta W \ll kT \ll W_1 - W_0$, (9.75) gives

$$P_E \simeq P_E^0 \Delta W/2kT \tag{9.77}$$

so that the mobility falls as the temperature is raised. With further increase in temperature higher pairs of levels begin to be occupied and the mobility may increase slightly because of the higher tunnelling probability (larger ΔW_n) for these levels, before it once again falls as T^{-1}. With this increase in temperature, however, the classical jump probability has been increasing and, at sufficiently high temperatures, its contribution will dominate the quantum behaviour, giving a mobility varying as $\exp(-W^t/kT)$. This overall behaviour is shown in fig. 9.13.

The quantum behaviour is important only if the quantum contribution to the jump rate P_E in (9.75) is greater than the classical contribution at the temperature considered. It is difficult to decide this point conclusively since the quantum jump rate depends critically upon the shape and width of the energy barrier, while the classical rate is determined primarily by the barrier height. Consider first an ion state, in which case the effective charge q should be replaced by p, defined in (9.24), throughout. For a classical

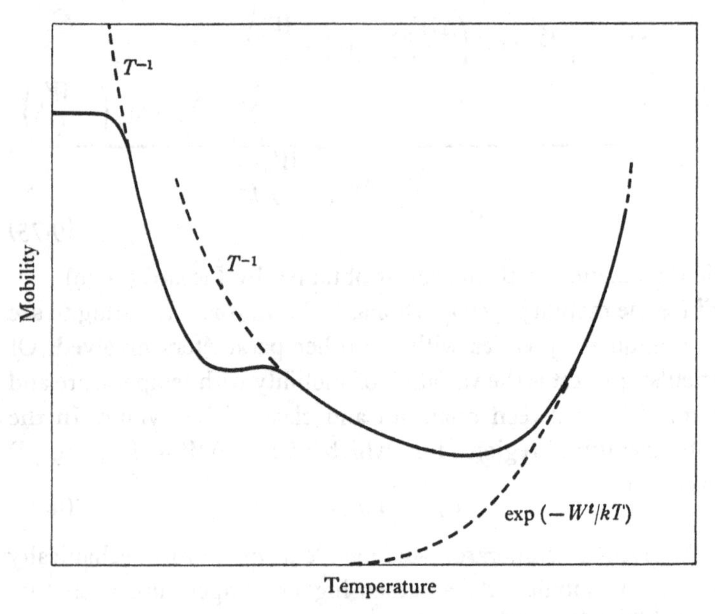

Fig. 9.13. Schematic behaviour of ion-state mobility as a function of temperature as calculated from equation (9.75).

oscillation frequency of $10^{14}\ s^{-1}$ for a proton in one of the wells, and using a one-dimensional approximation, the energy splitting ΔW of the ground state is of order 10^{-5} eV for an O–H...O bond length of 2·76 Å, as can be calculated from (9.66) using the known simple harmonic oscillator wave functions. This splitting increases by about a factor of 10 for every 0·1 Å decrease in the bond length (Baker, 1956; Weissmann & Cohan, 1965*b*). The energy level separation $h\nu$ is about 0·4 eV so that the ground state lies about 0·2 eV above the bottom of the wells. If the individual wells were exactly parabolic and met in a cusp, we should expect a barrier height of about 1·6 eV, which is close to the value 0·98 eV calculated quantum-mechanically by Weissmann & Cohan (1965*b*). The $n = 1$ states thus lie about halfway up the barrier maximum and the $n = 2$ states are probably above it. A fairly extensive parametric study of energy levels in such double wells has been published by Somorjai & Hornig (1962).

The quantum tunnelling rate $P_{A \leftrightarrow B}$ calculated from (9.68) is about 10^{10} s^{-1} and the mobility at -10 °C, found from (9.77) by noting that each quantum jump transfers the ion state to a neighbouring molecule, is about 10^{-4} cm^2 V^{-1} s^{-1}. If the energy barrier W^t is about 1 eV, then the classical component of the mobility calculated from (9.75) is extremely small, so that quantum tunnelling certainly determines the mobility. The estimated quantum mobility is, however, several orders of magnitude smaller than observed experimentally and this seems to require a modification to the calculation, as we shall see presently.

The smaller mobility of negative ion states is easily understood, since the potential barrier is significantly higher (see table 7.2). Similarly, the fact that deuteron ion-state mobility is less than that for the proton case by about a factor 7 (table 7.1) follows immediately from its larger mass, which gives a smaller value for ΔW in (9.66).

When we consider the motion of orientational defects, the situation is very different. The jump distance $2l$ is 1·65 Å compared with 0·76 Å for ion states so that the quantum tunnelling frequency is reduced by many orders of magnitude. At the same time the barrier height for classical jumping is very much lowered because only bond angle distortion rather than O–H bond stretching is required. A value of W^t near the experimentally observed 0·23 eV seems entirely reasonable. This reduction in barrier height completely outweighs the associated reduction in natural frequency ν, with the result that the calculated classical mobility is in moderately good agreement with the experimental value near 10^{-4} cm^2 V^{-1} s^{-1}.

Returning to consideration of ion states, we seek an explanation of the fact that the calculated quantum mobility is so much less than the observed mobility. In the first place, we assumed that the O–H...O distance from an ion to its nearest neighbours is the same as between two neutral water molecules, 2·76 Å. This is not likely to be the case, since the neighbours are attracted by the ionic charge, and the reduced jump distance will increase the mobility greatly. A decrease of bond length round the ion to about 2·6 Å would increase ΔW to about 10^{-3} eV and yield a calculated mobility in quite good agreement with experiment.

There is, however, another possibility which was first investigated by Sussmann (1964). The elementary tunnelling process which we have considered neglects the fact that each one-proton system is embedded in a crystal at finite temperature and thus may interact with lattice vibrations. These vibrations may be thought of quite simply as periodically modifying the O–O distances in the crystal and, from the amplitudes given in table 6.2, these changes may be up to $+0{\cdot}1$ Å. The perturbations which these vibrations introduce may be included in the Hamiltonian and, by (9.66), then contribute to ΔW and so, by (9.77), increase the mobility at finite temperatures.

Sussmann's treatment is not directly applicable to our present study, since he assumes fields which are large enough that $qEl > \Delta E$, which is not usually the case in conductivity experiments in ice. He does show, however, that the interactions can be analysed in terms of emission and absorption of phonons, a topic to which we shall return presently.

If we continue to think of the one-proton systems as forming part of the large crystal system then we are brought naturally to consider the relations between successive proton jumps. Our simple treatment assumed that these are completely independent but, as was first pointed out by Onsager & Dupuis (1960), this need not be the case. In fact it seems quite reasonable from momentum considerations that successive jumps may be correlated in such a way that the arrival of a proton at a particular molecular site encourages one of the other protons on the molecule to jump with the same general direction of motion. This is not thought of as happening in the motion of D- or L-defects because each proton makes very many vibrations in its well before jumping, but is a very real possibility for the much faster quantum-mechanically determined motion of ion states. Such a correlation would modify the normal Einstein relation between mobility and diffusion coefficient and, as we have seen, there is evidence that this is so.

If such a correlation exists to a significant extent, then it is useful to treat the motion of ion states in the same sort of way as the motion of an electron in the conduction band of a semiconductor is treated and this has been done in a one-dimensional approximation by Gosar & Pintar (1964), following an earlier treatment by Gosar (1963). Thus, instead of considering the state $\psi(n)$ in which the ion is localized on a molecule at a position $x = nr_0$, we deal with wave-like excitations $\psi(k)$ of the form

$$\psi(k) = \sum_n \exp(iknr_0)\,\psi(n), \tag{9.78}$$

where k is the wave vector of the state. The states of different k then form a band in which dependence of energy on k is given by

$$W(k) = -\Delta W \cos kr_0, \tag{9.79}$$

where ΔW is the tunnelling matrix element given by (9.66). The width of the band is then $2\Delta W$, which is only of the order of 10^{-3} eV when readjustment of bond lengths around the ion state is taken into account.

In this formalism the mobility is determined partly by the value of ΔW (which occurs as ΔW^2) and partly by the scattering of ion-state waves by lattice vibrations in the form of phonons. Because the ion-state band is so narrow and the difference between proton vibrational levels is so large compared with typical phonon energies, it is not possible for $\psi(k)$ to be scattered to a new state $\psi(k')$ by emission or absorption of a single phonon. Instead, so that energy and momentum can be conserved, scattering must occur by the simultaneous emission and absorption of a pair of phonons of nearly equal energy. The analysis is therefore rather complicated but, if we assume that orientational defects are present in sufficient concentration that polarization effects do not block ion paths, the calculated mobility is

$$\mu_+ = \frac{pr_0 l \Delta W^2 \tau}{\hbar^2 kT}, \tag{9.80}$$

where p is the effective polarization charge on an ion state, r_0 is the distance between molecules, $2l$ the distance between potential minima along a bond and τ is the relaxation time for scattering of ion states by phonons. There is some uncertainty in the calculation of τ and ΔW because of uncertainty in the wave functions used to calculate the matrix elements involved, but the likely values are

$\Delta W \simeq 1 \times 10^{-3}$ eV, $\tau \simeq 3 \times 10^{-12}$ s at -10 °C, which give fairly good agreement with the observed mobility. The ion-state mean free path is only about 8 Å, so that the average correlation in ion motion would appear to extend over about three individual jumps.

This result suggests the possibility of an even more extreme approach to the treatment of ion-state mobility in which the ion states are considered to move as essentially free particles which are then scattered by the lattice phonons. A semi-classical treatment along these lines by Kim & Schmidt (1967) gives reasonably good agreement with experiment. It is also worth noting that, from the band structure described by (9.79), a proton should behave, for small k, like a particle of effective mass m^*, where m^* is given by the familiar formula for the electronic case

$$m^* = \hbar^2 \left(\frac{\partial^2 W}{\partial k^2}\right)^{-1} = \hbar^2/(r_0^2 \Delta W). \tag{9.81}$$

If $\Delta W = 1 \times 10^{-3}$ eV as suggested, then m^* turns out to be about half the normal proton mass, which seems reasonable.

Further insight into the process of ion-state motion can be obtained from recent studies of the protonic Hall effect in ice by Bullemer & Riehl (1968). As in the ordinary Hall effect, electric and magnetic fields are applied at right angles and the transverse current or voltage is measured, great care being necessary because of surface conductivity and other extraneous effects. In pure ice the Hall coefficient is positive, indicating that $\mu_+ > \mu_-$ as we have already seen. The magnitude of the Hall coefficient at -8 °C, 1.5×10^9 cm^3 A^{-1} s^{-1}, leads to the carrier concentration and mobility values

$$n_+^{\mathrm{H}} = 0.4 \times 10^{10}\ \mathrm{cm}^{-3}, \quad \mu_+^{\mathrm{H}} = 1.4\ \mathrm{cm}^2\,\mathrm{V}^{-1}\,\mathrm{s}^{-1}, \tag{9.82}$$

which differ significantly from those found by other methods, μ_+^{H} in particular being 20 times as large as the drift mobility measured by Eigen *et al.* (1964). This does not, however, really imply any physical disagreement, since the Hall mobility is physically different from the drift mobility. Suppose, for example, that the motion of an ion state is such that it tunnels freely for a time τ_s and is then trapped for a time τ_t before beginning to move again. Then these times enter the averages for drift mobility and Hall mobility differently so that

$$\frac{\mu_+^{\mathrm{H}}}{\mu_+^{\mathrm{D}}} = \frac{2(\tau_s + \tau_t)m_{\mathrm{D}}^*}{3\tau_s m_{\mathrm{H}}^*}, \tag{9.83}$$

where the possibility has been allowed that the effective mass m^* may also be different for Hall and drift motions. Clearly if τ_t is large compared with τ_s then μ_+^{H} will be greater than μ_+^{D}. There is a somewhat similar effect in measurement of n_+^{H}.

The Hall effect measurements also show that the mobility increases as the temperature is lowered, a trend predicted by the quantum relations (9.76) or (9.80) but completely opposite to the classical behaviour (9.62). This trend had also been noted previously in low-temperature high-field studies by Engelhardt & Riehl (1966), where also there was evidence of proton trapping.

From the discussion above it is clear that we now have a reasonably good understanding of the details of proton motion in ice crystals, but it is equally clear that there are many more subtle effects which still require further detailed study.

9.8. Miscellaneous effects

The electrical properties of ice are so varied and have been studied so extensively that we cannot hope, in a book of this type, even to list them all, much less give any detailed treatment. There are, however, one or two studies of particular interest which warrant brief mention.

First consider the surface conductivity of ice. This is not important at temperatures below about -20 °C but becomes rapidly larger as the melting point is approached, so much so that it may dominate the behaviour of thin samples above -10 °C. This phenomenon has been studied in some detail by Jaccard (1967), who concludes that at -11 °C the surface conductivity is $10^{-10}\Omega^{-1}$, while Bullemer & Riehl (1966, 1968) found that the surface conductivity above this temperature has the very large apparent activation energy of 1.4 eV. This behaviour may be explained by the speculative theory of Fletcher (1968) which, on the basis of molecular, electrostatic and thermodynamic arguments, concludes that, at temperatures within about 10 degC of the melting point, the surface of ice is covered by a thin quasi-liquid disordered layer. The concentration of H_3O^+ and OH^- ions in this surface layer is calculated to be very high, because of the potential gradient due to molecular orientation, and the film thickness, which is calculated to be about 40 Å at -5 °C, increases rapidly as the temperature nears the melting point, thus giving at least a qualitative explanation of the phenomenon.

Another set of phenomena of interest is that connected with charge separation and static electrification. There have been many studies of the effects of friction between ice surfaces and of evaporation as charging mechanisms which we will not discuss here, but certainly one of the most striking phenomena is the electrification produced by the freezing of dilute aqueous solutions. This effect was first noted by the Brazilian physicist Costa Ribeiro for a variety of melts, and sometimes bears his name, but is more usually associated with Workman & Reynolds (1950), who did the initial work on aqueous solutions and proposed its connexion with thunderstorm electrification. A recent survey has been given by Pruppacher *et al.* (1968), who also made a systematic study of the effects of different ionic solutes. Briefly, the experimental facts are that, if water containing a small amount of ionic solute is allowed to freeze, then a potential difference of some tens of volts may be developed between the ice and the solution and currents of the order of 10^{-7} A may be sustained by growing interfaces only a few square centimetres in area. The developed potential does not depend greatly upon freezing rate but does depend systematically on the nature of the solute. The potential is greatest for a solute concentration of about 10^{-4} mole l^{-1} and becomes essentially zero for concentrations greater than 10^{-2} or less than 10^{-6} mole l^{-1}.

The explanation of the effect apparently lies in the selective incorporation of ions of one sign into the ice structure during freezing. For alkali halide solutions the ice is negative with respect to the solution and the potential difference is greatest for fluorides and least for iodides, depending very little upon the cation involved. This is in accord with the generally high solubility of the fluoride ion in ice. In the case of ammonium salts, ammonium fluoride makes the ice negative while the other halides give it a positive potential, which is again reasonable because of the large solubility of the ammonium ion.

Finally, let us return to static properties and consider the possible existence of a ferroelectric transition in ice. The statistical model for molecular orientations in ice, which we discussed in chapter 2, is based upon the assumption that the energy of all dipole configurations is essentially the same. This is unlikely to be exactly true and it is really sufficient for the model to require that the energy differences between possible configurations are small compared with kT at a temperature at which the orientational relaxation time is comparable to the time involved in the experiment. Specific heat measurements and the agreement between measured and calculated residual entropies would seem to show this to be so. If, however, there is a configuration of lowest energy, and if this configuration gives a net polarization to the structure, then we might expect ice to exhibit some sort of ferroelectric transition at low temperatures, provided the time scale or some other feature of the experiment could be adjusted to allow the necessary equilibrium to be attained. The agreement of the experimental value of the residual entropy with that predicted by the Pauling model implies that the normal slow cooling used in heat capacity studies does not allow the transition to take place.

The normal theory of ferroelectric transitions (Slater, 1941) gives little help on this point, since the energy difference involved is of subtle origin, rather than depending on nearest-neighbour interactions. Extrapolation of measured static dielectric constants to low temperatures might perhaps be expected to show a $(T-T_C)^{-1}$ dependence rather than T^{-1} if there is a ferroelectric Curie point at T_C, but the extrapolations of different sets of data are in disagreement even as to the sign of the formal T_C. Direct measurements by Dengel *et al.* (1964) of the integrated current needed at a given temperature to charge a capacitor having an ice dielectric suggest, however, that the dielectric constant of ice goes through a sharp maximum near 100 °K, indicating a ferroelectric transition. This is supported by measurements of depolarization currents in slowly warmed samples. This behaviour, however, could only be observed with ice which was not extremely pure and, though this may only mean that impurities are necessary to give reasonably short relaxation times, it does leave the possibility that the effect is not intrinsic to ice but is associated directly with impurity effects. Somewhat similar results have been reported by Cubiotti & Geracitano (1967) for Ice I_c but have not yet been confirmed.

Despite this uncertainty about the existence of a ferroelectric transition, it is possible to produce an ice crystal with a permanent macroscopic dipole moment—an electret—by polarizing the crystal at about -15 °C and then cooling it to liquid-air temperature (Gelin & Stubbs, 1965). These electrets are essentially stable at these low temperatures but discharge slowly at higher temperatures, the decay half-life being of the order of 1 min at -60 °C, though the decay is not exponential.

9.9. Conclusion

Because this book did not set out to be a catalogue of the properties of ice but rather to give a connected view of some of its most important attributes, there remain many things which have not even been mentioned. We have, however, seen enough to realize that most of these properties depend critically upon the structure of the water molecule, for this molecule, remaining intact through almost all processes, dictates the possible crystal structures for ice and the ways in which they can vibrate, deform or rearrange.

It is true that many of the most interesting properties are found to depend upon impurities or imperfections and indeed this is fairly generally true in solid-state physics, but these imperfections all act out their roles against the background of the perfect crystal.

Water and ice are important and unique materials whose properties show at once striking similarities to and striking differences from those of other substances. The work we have described has led to a coherent and useful appreciation of the reasons behind many observed phenomena but the field is so rich that it is very far from being completely explored, let alone exhausted.

REFERENCES

[*Numbers in brackets following each reference indicate the text pages where the reference has been cited.*]

Abraham, F. F. & Pound, G. M. (1968). A re-examination of homogeneous nucleation theory: Statistical thermodynamics aspects. *J. Chem. Phys.* **48**, 732–40. [88]

Auty, R. P. & Cole, R. H. (1952). Dielectric properties of ice and solid D_2O. *J. Chem. Phys.* **20**, 1309–14. [201–2, 209–10]

Bader, R. F. W. & Jones, G. A. (1963). The electron density distribution in hydride molecules. I. The water molecule. *Can. J. Chem.* **41**, 586–606. [10]

Baker, A. N. (1956). Polarizability of the hydrogen bridge. *J. Chem. Phys.* **25**, 381. [238]

Banyard, K. E. (1960). Diamagnetism as a test of wave functions for some simple molecules. *J. Chem. Phys.* **33**, 832–6. [12]

Banyard, K. E. & March, N. H. (1957). Distribution of electrons in the water molecule. *J. Chem. Phys.* **26**, 1416–20. [12]

Barker, J. A. (1963). *Lattice Theories of the Liquid State*. Oxford: Pergamon Press. [73]

Barnaal, D. E. & Lowe, I. L. (1967). Experimental free-induction-decay shapes and theoretical second moments for hydrogen in hexagonal ice. *J. Chem. Phys.* **46**, 4800–9. [33]

Barnes, W. J. (1929). The crystal structure of ice between 0 °C and −183 °C. *Proc. Roy. Soc. Lond.* A **125**, 670–93. [24, 28]

Bass, R. (1958*a*). Zur Theorie der mechanischen Relaxation des Eises. *Z. Phys.* **153**, 16–37. [176, 179]

Bass, R. (1958*b*). A theoretical analysis of the mechanical relaxation of single-crystalline ice. *Proc. Roy. Soc. Lond.* A **247**, 462–4. [179–80]

Bass, R., Rossberg, D. & Ziegler, G. (1957). Die elastischen Konstanten des Eises. *Z. Phys.* **149**, 199–203. [169–72]

Beaumont, R. H., Chihara, H. & Morrison, J. A. (1961). Transitions between different forms of ice. *J. Chem. Phys.* **34**, 1456–7. [59, 60]

Becker, R. & Döring, W. (1935). Kinetische Behandlung der Keimbildung in übersättigen Dampfen. *Ann. Phys., Lpz.* **24**, 719–52. [87]

Bernal, J. D. & Fowler, R. H. (1933). A theory of water and ionic solution, with particular reference to hydrogen and hydroxyl ions. *J. Chem. Phys.* **1**, 515–48. [10, 29, 42, 44, 80]

Bertie, J. E., Calvert, L. D. & Whalley, E. (1963). Transformations of Ice II, Ice III and Ice V at atmospheric pressure. *J. Chem. Phys.* **38**, 840–6. [52, 58, 64, 66]

Bertie, J. E., Calvert, L. D. & Whalley, E. (1964). Transformations of Ice VI and Ice VII at atmospheric pressure. *Can. J. Chem.* **42**, 1373–8. [52, 62, 68, 71]

Bertie, J. E. & Whalley, E. (1964*a*). Infrared spectra of Ices I_h and I_c in the range 4000 to 350 cm^{-1}. *J. Chem. Phys.* **40**, 1637–45. [58, 136–7]

Bertie, J. E. & Whalley, E. (1964*b*). Infrared spectra of Ices II, III and V in the range 4000 to 350 cm^{-1}. *J. Chem. Phys.* **40**, 1646–59. [61–2, 64, 66]

Bertie, J. E. & Whalley, E. (1967). Optical spectra of orientationally disordered crystals. II. Infrared spectrum of Ice I_h and Ice I_c from 360 to 50 cm^{-1}. *J. Chem. Phys.* **46**, 1271–84. [136–8, 140]

Bigg, E. K. (1953). The supercooling of water. *Proc. Phys. Soc.* B **66**, 688–94. [86]

Bishop, D. M., Hoyland, J. R. & Parr, R. G. (1963). Simple one-centre calculation of breathing force constants and equilibrium internuclear distances for NH_3, H_2O and HF. *Molec. Phys.* **6**, 467–76. [3]

Bishop, D. M. & Randić, M. (1966). A theoretical investigation of the water molecule. *Molec. Phys.* **10**, 517–28. [3, 8]

Bjerrum, N. (1951). Structure and properties of ice. *K. danske Vidensk. Selsk. Mat.-fis. Medd.* **27** (1), 3–56. [10, 35, 42, 45, 141, 149, 153, 210, 212, 223]

Bjerrum, N. (1952). Structure and properties of ice. *Science, N.Y.* **115**, 385–90. [193]

Blackman, M. & Lisgarten, N. D. (1958). Electron diffraction investigations into the cubic and other structural forms of ice. *Adv. Phys.* **7**, 189–98. [58]

Blicks, H., Dengel, O. & Riehl, N. (1966). Diffusion von Protonen (Tritonen) in reinen und dotierten Eis-Einkristallen. *Phys. kondens. Materie* **4**, 375–81. [160]

Bloembergen, N., Purcell, E. M. & Pound, R. V. (1948). Relaxation effects in nuclear magnetic resonance absorption. *Phys. Rev.* **73**, 679–712. [33]

Blue, R. W. (1954). The librational heat capacity of ice and heavy ice. *J. Chem. Phys.* **22**, 280–3. [141]

Bogorodskii, V. V. (1964). Elastic moduli of ice crystals. *Soviet Physics Acoustics* **10**, 124–6. [169–70]

Bradley, R. S. (1957). The electrical conductivity of ice. *Trans. Faraday Soc.* **53**, 687–91. [212]

Bradley, R. S. (ed.) (1963). *High Pressure Physics and Chemistry*. London: Academic Press. [50]

Brady, G. W. & Romanow, W. J. (1960). Structure of water. *J. Chem. Phys.* **32**, 306. [74]

Brice, J. C. (1965). *The Growth of Crystals from the Melt*. Amsterdam: North-Holland. [118]

Bridgman, P. W. (1912). Water in the liquid and five solid forms, under pressure. *Proc. Am. Acad. Arts Sci.* **47**, 441–558. [50, 61]

Bridgman, P. W. (1935). The pressure-volume-temperature relations of the liquid and the phase diagram of heavy water. *J. Chem. Phys.* **3**, 597–605. [53–5, 66]

Bridgman, P. W. (1937). The phase diagram of water to 45,000 kg/cm^2. *J. Chem. Phys.* **5**, 964–6. [50, 54]

Bridgman, P. W. (1952). *The Physics of High Pressure*. London: G. Bell and Sons. [50]

Brown, A. J. & Whalley, E. (1966). Preliminary investigation of the phase boundaries between Ice VI and VII and Ice VI and VIII. *J. Chem. Phys.* **45**, 4360–1. [50, 54]

Brownscombe, J. L. & Mason, B. J. (1966). Measurement of the thermoelectric power of ice by an induction method. *Phil. Mag.* **14**, 1037–47. [227–8, 230]

Bryant, G. W. (1967). Thermoelectric power of single crystals of ice containing HF or NH_3. *Phil. Mag.* **16**, 495–504. [227–8, 230]

Bryant, G. W. & Fletcher, N. H. (1965). Thermoelectric power of ice containing HF or NH_3. *Phil. Mag.* **12**, 165–76. [227–8]

Bryant, G. W., Hallett, J. & Mason, B. J. (1960). The epitaxial growth of ice on single crystalline substrates. *J. Phys. Chem. Solids* **12**, 189–95. [101, 123]

Bryant, G. W. & Mason, B. J. (1960). Etch pits and dislocations in ice crystals. *Phil. Mag.* **5**, 1221–7. [193]

Buckingham, A. D. (1965). Theory of long-range dispersion forces. *Disc. Faraday Soc.* **40**, 232–8. [19, 21]

Buijs, K. & Choppin, G. R. (1963). Near-infrared studies of the structure of water. I. Pure water. *J. Chem. Phys.* **39**, 2035–41. [82]

Bullemer, B. & Riehl, N. (1966). Bulk and surface conductivity of ice. *Solid State Commun.* **4**, 447–8. [243]

Bullemer, B. & Riehl, N. (1968). Hall-Effekt in Protonen in Eis. *Phys. kondens. Materie* **7**, 248–60. [242–3]

Burton, W. K., Cabrera, N. & Frank, F. C. (1951). The growth of crystals and the equilibrium structure of their surfaces. *Phil. Trans. Roy. Soc.* **243**, 299–358. [105, 107–8, 124]

Butkovich, T. R. (1959). Thermal expansion of ice. *J. Appl. Phys.* **30**, 350–3. [131]

Cahn, J. W. (1960). Theory of crystal growth and interface motion in crystalline materials. *Acta Metall.* **8**, 554–62. [105, 108]

Cahn, J. W., Hillig, W. B. & Sears, G. W. (1964). The molecular mechanism of solidification. *Acta Metall.* **12**, 1421–39. [105]

Camp, P. R. & Creamer, J. (1966). Rate of growth of ice at water-metal interfaces. *J. Chem. Phys.* **45**, 2709–10. [116]

Campbell, E. S., Gelernter, G., Heinen, H. & Moorti, V. R. G. (1967). Interpretation of the energy of hydrogen bonding. *J. Chem. Phys.* **46**, 2690–707. [42, 47]

Cassel, E. J. (1935). Ultraviolet absorption of ice. *Proc. Roy. Soc. Lond.* A **153**, 534–41. [212]

Chan, R. K., Davidson, D. W. & Whalley, E. (1965). Effect of pressure on the dielectric properties of Ice I. *J. Chem. Phys.* **43**, 2376–83. [211–12, 217]

Chaudhuri, A. R., Patel, J. R. & Rubin, L. G. (1962). Velocities and densities of dislocations in germanium and other semiconductor crystals. *J. Appl. Phys.* **33**, 2736–46; (1963) **34**, 240. [194–5]

Chidambaram, R. (1961). A bent hydrogen bond model for the structure of Ice I. *Acta Cryst.* **14**, 467–8. [42]

Cohan, N. V., Cotti, M., Iribarne, J. V. & Weissmann, M. (1962). Electrostatic energies in ice and the formation of defects. *Trans. Faraday Soc.* **58**, 490–8. [155]

Cole, K. S. & Cole, R. H. (1941). Dispersion and absorption in dielectrics. *J. Chem. Phys.* **9**, 341–51. [208]

Collins, J. G. & White, G. K. (1964). Thermal expansion of solids. *Prog. Low Temp. Phys.* **4**, 450–79. [132, 134]

Conway, B. E., Bockris, J. O'M. & Linton, H. (1956). Proton conductance and the existence of the H_3O^+ ion. *J. Chem. Phys.* **24**, 834–50. [234]

Cottrell, A. H. (1953). *Dislocations and Plastic Flow in Crystals*. Oxford University Press. [190]

Coulson, C. A. (1951). Critical survey of the method of ionic-homopolar resonance. *Proc. Roy. Soc. Lond.* A **207**, 63–73. [9]

Coulson, C. A. (1959). The hydrogen bond. In *Hydrogen Bonding* (ed. D. Hadzi), pp. 339–60. London: Pergamon Press. [42]

Coulson, C. A. (1961). *Valence*. Oxford University Press. [1, 235]

Coulson, C. A. & Eisenberg, D. (1966*a*). Interactions of H_2O molecules in Ice I. The dipole moment of an H_2O molecule in ice. *Proc. Roy. Soc. Lond.* A **291**, 445–53. [19, 46]

Coulson, C. A. & Eisenberg, D. (1966*b*). Interactions of H_2O molecules in ice. II. Interaction energies of H_2O molecules in ice. *Proc. Roy. Soc. Lond.* A **291**, 454–9. [42, 46]

Cross, P. C., Burnham, J. & Leighton, P. A. (1937). The Raman spectrum and the structure of water. *J. Am. Chem. Soc.* **59**, 1134–47. [82]

Cubiotti, G. & Geracitano, R. (1967). Ferroelectric behaviour of cubic ice. *Phys. Letters*, A **24a**, 179–80. [246]

Danford, M. D. & Levy, H. A. (1962). The structure of water at room temperature. *J. Am. Chem. Soc.* **84**, 3965–6. [74]

Dantl, G. (1962). Wärmeausdehnung von H_2O- und D_2O-Einkristallen. *Z. Phys.* **166**, 115–18. [131–2]

Davis, C. M. & Litovitz, T. A. (1965). Two-state theory of the structure of water. *J. Chem. Phys.* **42**, 2563–76. [83]

Dean, J. W. & Timmerhaus, K. D. (1963). Thermal conductivity of solid H_2O and D_2O at low temperatures. *Adv. Cryogen. Engng* **8**, 263–7. [143–4]

Debye, P. (1929). *Polar Molecules*. Reprinted 1945. New York: Dover. [20, 207]

Decroly, J. C., Gränicher, H. & Jaccard, C. (1957). Caractère de la conductivité électrique de la glace. *Helv. Phys. Acta* **30**, 465–7. [212]

Decroly, J. C. & Jaccard, C. (1957). Croissance et raffinage contrôlés de cristaux de glace. *Helv. Phys. Acta* **30**, 468–9. [119]

Delibaltas, P., Dengel, O., Helmreich, D., Riehl, N. & Simon, H. (1966). Diffusion von ^{18}O in Eis-Einkristallen. *Phys. kondens. Materie* **5**, 166–70. [159]

Dengel, O., Eckener, U., Plitz, H. & Riehl, N. (1964). Ferro-electric behaviour of ice. *Phys. Letters* **9**, 291–2. [245]

Dengel, O., Jacobs, E. & Riehl, N. (1966). Diffusion von Tritonen in NH_4F-dotierten Eis-Einkristallen. *Phys. kondens. Materie* **5**, 58–9. [160]

Dengel, O. & Riehl, N. (1963). Diffusion von Protonen (Tritonen) in Eiskristallen. *Phys. kondens. Materie* **1**, 191–6. [160]

Dengel, O., Riehl, N. & Schleippmann, A. (1966). Messungen der Dielektrizitätskonstanten von reinem und NH_4F-dotiertem Eis. *Phys. kondens. Materie* **5**, 83–8. [224]

Di Marzio, E. A. & Stillinger, F. H. (1964). Residual entropy of ice. *J. Chem. Phys.* **40**, 1577–93. [37]

Dorsey, N. E. (1940). *Properties of Ordinary Water-substance*. New York: Reinhold Pub. Corp. [ix]

Dougherty, T. J. (1965). Electrical properties of ice. I. Dielectric relaxation in pure ice. *J. Chem. Phys.* **43**, 3247–52. [218]

Dowell, L. G. & Rinfret, A. P. (1960). Low-temperature forms of ice as studied by X-ray diffraction. *Nature, Lond.* **188**, 1144–8. [58, 60]

Duncan, A. B. F. & Pople, J. A. (1953). The structure of some simple molecules with lone pair electrons. *Trans. Faraday Soc.* **49**, 217–24. [9]

Dunitz, J. D. (1963). Nature of orientational defects in ice. *Nature, Lond.* **197**, 860–2. [155]

Durand, M., Deleplanque, M. & Kahane, A. (1967). Bulk conductivity of ice between −25 and −100 °C with ion exchange membranes. *Solid State Commun.* **5**, 759–60. [212]

Edwards, G. R. & Evans, L. F. (1962). Effect of surface charge on ice nucleation by silver iodide. *Trans. Faraday Soc.* **58**, 1649–55. [102]

Edwards, G. R., Evans, L. F. & La Mer, V. K. (1962). Ice nucleation by monodisperse silver iodide particles. *J. Colloid Sci.* **17**, 749–58. [100]

Eigen, M. & De Maeyer, L. (1958). Self-dissociation and protonic charge transport in water and ice. *Proc. Roy. Soc. Lond.* A **247**, 505–33. [148–51]

Eigen, M., De Maeyer, L. & Spatz, H. C. (1964). Über das kinetische Verhalten von Protonen und Deuteronen in Eiskristallen. *Berichte der Bunsengesellschaft* **68**, 19–29. [150–1, 212, 218, 225, 242]

Eisenberg, D. & Coulson, C. A. (1963). Energy of formation of D-defects in ice. *Nature, Lond.* **199**, 368–9. [155]

Ellison, F. O. & Shull, H. (1953). An LCAO MO self-consistent field calculation of the ground state of H_2O. *J. Chem. Phys.* **21**, 1420–1 [5]

Ellison, F. O. & Shull, H. (1955). Molecular Calculations. I. LCAO MO self-consistent field treatment of the ground state of H_2O. *J. Chem. Phys.* **23**, 2348–57. [5, 7]

Engelhardt, H. & Riehl, N. (1966). Zur protonischen Leitfähigkeit von Eis-Einkristallen bei tiefen Temperaturen und hohen Feldstärken. *Phys. kondens. Materie* **5**, 73–82. [243]

Eshelby, J. D. (1961). Dislocations in visco-elastic materials. *Phil. Mag.* **6**, 953–63. [194]

Evans, L. F. (1967*a*). Ice nucleation under pressure and in salt solution. *Trans. Faraday Soc.* **63**, 3060–71. [101]

Evans, L. F. (1967*b*). Selective nucleation of the high pressure ices. *J. Appl. Phys.* **38**, 4930–2. [103]

Eyring, H. (1935). The activated complex in chemical reactions. *J. Chem. Phys.* **3**, 107–15. [92]

Falk, M. & Ford, T. A. (1966). Infrared spectrum and structure of liquid water. *Can. J. Chem.* **44**, 1699–1707. [76, 79, 82]
Fletcher, N. H. (1958). Size effect in heterogeneous nucleation. *J. Chem. Phys.* **29**, 572–6, **31**, 1136–7. [98, 100]
Fletcher, N. H. (1959). Entropy effect in ice crystal nucleation. *J. Chem. Phys.* **30**, 1476–82. [102]
Fletcher, N. H. (1962*a*). *The Physics of Rainclouds*. Cambridge University Press. [86, 102]
Fletcher, N. H. (1962*b*). Surface structure of water and ice. *Phil. Mag.* **7**, 255–69; (1963) **8**, 1425–6. [126]
Fletcher, N. H. (1963). Nucleation by crystalline particles. *J. Chem. Phys.* **38**, 237–40. [99]
Fletcher, N. H. (1964). Crystal interfaces. *J. Appl. Phys.* **35**, 234–40. [101]
Fletcher, N. H. (1967). Structure and energy of crystal interfaces. II. A simple explicit calculation. *Phil. Mag.* **16**, 159–64. [101]
Fletcher, N. H. (1968). Surface structure of water and ice. II. A revised model. *Phil. Mag.* **18**, 1287–1300. [126–7, 243]
Fletcher, N. H. & Adamson, P. L. (1966). Structure and energy of crystal interfaces. I. Formal development. *Phil. Mag.* **14**, 99–110. [101]
Flubacher, P., Leadbetter, A. J. & Morrison, J. A. (1960). Heat capacity of ice at low temperatures. *J. Chem. Phys.* **33**, 1751–5. [35, 134–6]
Frank, F. C. (1949). The influence of dislocations on crystal growth. *Disc. Faraday Soc.* **5**, 48–54. [106]
Frank, H. S. (1958). Covalency in the hydrogen bond and the properties of water and ice. *Proc. Roy. Soc.* A **247**, 481–92. [42, 78]
Frank, H. S. & Evans, M. W. (1945). Free volume and entropy in condensed systems. III. *J. Chem. Phys.* **13**, 507–32. [97]
Frank, H. S. & Quist, A. S. (1961). Pauling's model and the thermodynamic properties of water. *J. Chem. Phys.* **34**, 604–11. [84]
Frank, H. S. & Wen, W-Y. (1957). Structural aspects of ion-solvent interaction in aqueous solutions: a suggested picture of water structure. *Disc. Faraday Soc.* **24**, 133–40. [78, 85]
Frenkel, J. (1939). A general theory of heterophase fluctuations and pretransition phenomena. *J. Chem. Phys.* **7**, 538–47. [87]
Frenkel, J. (1946). *Kinetic Theory of Liquids*. Oxford: Clarendon Press. Ch. 7. [87, 93]
Fröhlich, H. (1949). *Theory of Dielectrics*. Oxford: Clarendon Press. [201, 205]
Fukuta, N. (1958). Experimental investigations on the ice forming ability of various chemical substances. *J. Meteorol.* **15**, 17–26. [102]
Fukuta, N. (1966). Experimental studies of organic ice nuclei. *J. Atmos. Sci.* **23**, 191–6. [102]
Furashov, N. I. (1966). Far infrared absorption by atmospheric water vapour. *Optics Spectrosc.* **20**, 234–7. [17]
Furukawa, K. (1962). The radial distribution curves of liquids by diffraction methods. *Rep. Prog. Phys.* **25**, 395–440. [73]
Garten, V. A. & Head, R. B. (1965). A theoretical basis of ice nucleation by organic crystals. *Nature, Lond.* **205**, 160–2. [102]

Gelin, H. & Stubbs, R. (1965). Ice electrets. *J. Chem. Phys.* **42**, 967–71. [246]

Ghormley, J. A. (1956). Thermal behaviour of amorphous ice. *J. Chem. Phys.* **25**, 599. [60]

Ghormley, J. A. (1968). Enthalpy changes and heat-capacity changes in the transformations from high-surface-area amorphous ice to stable hexagonal ice. *J. Chem. Phys.* **48**, 503–8. [60]

Giaque, W. F. & Stout, J. W. (1936). The entropy of water and the third law of thermodynamics. The heat capacity of ice from 15 to 273° K. *J. Am. Chem. Soc.* **58**, 1144–50. [34, 134–5]

Glaeser, R. M. & Coulson, C. A. (1965). Multipole moments of the water molecule. *Trans. Faraday Soc.* **61**, 389–91. [8–10, 47]

Glen, J. W. (1955). The creep of polycrystalline ice. *Proc. Roy. Soc. Lond.* A **228**, 519–38. [197]

Glen, J. W. (1958). The mechanical properties of ice. *Adv. Phys.* **7**, 254–65. [186]

Glen, J. W. (1967). The physics of the flow of glaciers. *Bull. Inst. Phys., Lond.* **18**, 135–40. [196]

Glen, J. W. (1968). The effect of hydrogen disorder on dislocation movement and plastic deformation of ice. *Phys. kondens. Materie* **7**, 43–51. [193]

Glen, J. W. & Jones, S. J. (1967). The deformation of ice single crystals at low temperatures. In *Physics of Snow and Ice* (ed. H. Ôura), pp. 267–75. Japan: Institute of Low Temperature Science, Hokkaido University. [188]

Glen, J. W. & Perutz, M. H. (1954). The growth and deformation of ice crystals. *J. Glaciol.* **2**, 397–403. [195]

Gold, L. W. (1958). Some observations on the dependence of strain on stress in ice. *Can. J. Phys.* **36**, 1265–75. [169, 172–3]

Gosar, P. (1963). On the mobility of the H_3O^+ ion in ice crystals. *Nuovo Cim.* **30**, 931–46. [241]

Gosar, P. & Pintar, M. (1964). H_3O^+ ion energy bands in ice crystals. *Phys. Stat. Solidi* **4**, 675–83. [241]

Gränicher, H. (1963). Properties and lattice imperfections of ice crystals and the behaviour of H_2O-HF solid solutions. *Phys. kondens. Materie* **1**, 1–12. [212]

Gränicher, H., Jaccard, C., Scherrer, P. & Steinemann, A. (1957). Dielectric relaxation and the electrical conductivity of ice crystals. *Disc. Faraday Soc.* **23**, 50–62. [212, 220]

Green, H. S. (1960). The structure of liquids. *Handb. Phys.* **10**, 1–133. [73]

Green, R. E. & Mackinnon, L. (1956). Determination of the elastic constants of ice single crystals by an ultrasonic pulse method. *J. Acoust. Soc. Am.* **28**, 1292. [169–70]

Grossweiner, L. I. & Matheson, M. S. (1954). Fluorescence and thermoluminescence of ice. *J. Chem. Phys.* **22**, 1514–26. [164]

Haas, C. (1962). On diffusion, relaxation and defects in ice. *Phys. Letters* **3**, 126–8. [160]

Hackforth, H. L. (1960). *Infrared Radiation.* New York: McGraw-Hill. ch. 4. [17, 18]

Hallett, J. (1961). The growth of ice crystals on freshly cleaved covellite surfaces. *Phil. Mag.* **6**, 1073–87. [123–4]

Hallett, J. (1963). The temperature dependence of the viscosity of supercooled water. *Proc. Phys. Soc.* **82**, 1046–50. [94]

Hallett, J. (1964). Experimental studies of the crystallization of supercooled water. *J. Atmos. Sci.* **21**, 671–82. [114]

Hallett, J. & Mason, B. J. (1958). The influence of temperature and supersaturation on the habit of ice crystals grown from the vapour. *Proc. Roy. Soc. Lond.* A, **247**, 440–53. [120–1]

Harrison, J. D. & Tiller, W. A. (1963). Ice interface morphology and texture developed during freezing. *J. Appl. Phys.* **34**, 3349–55. [111, 113]

Harrison, J. F. (1967). Some one-electron properties of H_2O and NH_3. *J. Chem. Phys.* **47**, 2990–6. [12]

Hasted, J. B. (1961). The dielectric properties of water. *Prog. Dielect.* **3**, 101–49. [76]

Herzeberg, G. (1945). *Molecular Spectra and Molecular Structure.* Vol. II. *Infrared and Raman Spectra of Polyatomic Molecules.* Princeton N.J.: Van Nostrand. [1, 16, 17]

Herzberg, G. (1966). *Molecular Spectra and Molecular Structure.* Vol. III. *Electronic Spectra and Electronic Structure of Polyatomic Molecules.* Princeton N.J.: Van Nostrand. [1]

Higashi, A., Koinuma, S. & Mae, S. (1964). Plastic yielding in ice single crystals. *Jap. J. Appl. Phys.* **3**, 610–16. [188–90]

Higashi, A., Koinuma, S. & Mae, S. (1965). Bending creep of ice single crystals. *Japan. J. Appl. Phys.* **4**, 575–82. [187]

Higashi, A. & Sakai, N. (1961*a*). Movement of small angle boundary of ice crystal. *J. Phys. Soc. Japan* **16**, 2359–60. [192]

Higashi, A. & Sakai, N. (1961*b*). Movement of small angle boundary of ice crystal. *J. Fac. Sci. Hokkaido Univ.* ser. II, **5**, 221–37. [192, 194]

Higuchi, K. (1958). The etching of ice crystals. *Acta Metall.* **6**, 636–42. [193]

Hillig, W. (1958). The kinetics of freezing of ice in the direction perpendicular to the basal plane. *Growth and Perfection of Crystals* (ed. R. H. Doremus, S. W. Roberts and D. Turnbull), pp. 350–60. New York: Wiley. [114, 117]

Hillig, W. & Turnbull, D. (1956). Theory of crystal growth in undercooled pure liquids. *J. Chem. Phys.* **24**, 914. [114]

Hirth, J. P. & Pound, G. M. (1963). Condensation and evaporation. *Prog. Materials Sci.* **11**. [88]

Hobbs, M. E., Jhon, M. S. & Eyring, H. (1966). The dielectric constant of liquid water and various forms of ice according to significant structure theory. *Proc. Natn. Acad. Sci. U.S.A.* **56**, 31–8. [206]

Hobbs, P. V. & Scott, W. D. (1965). Step growth on single crystals of ice. *Phil. Mag.* **11**, 1083–6. [126]

Hollins, G. T. (1964). Configurational statistics and the dielectric constant of ice. *Proc. Phys. Soc.* **84**, 1001–16. [36]

Hollomon, J. H. & Turnbull, D. (1953). Nucleation. In *Prog. Metal Phys.* **4**, 333–88. [87]

Honjo, G. & Shimaoka, K. (1957). Determination of hydrogen positions in cubic ice by electron diffraction. *Acta Cryst.* **10**, 710–11. [58]

Humbel, F., Jona, F. & Scherrer, P. (1953*a*). Anisotropie der Dielektrizitätskonstante des Eises. *Helv. Phys. Acta* **26**, 17–32. [201–2, 209–10]

Humbel, F., Jona, F. & Scherrer, P. (1953*b*). Le comportement dielectrique et les modules d'élasticité de monocristaux de glace. *J. Chim. Phys.* **50**, C40–3. [209]

International Union of Crystallography (1952). *International Tables for X-ray Crystallography*, vol. I. Birmingham: Kynoch Press. [24–5]

Iribarne, J. V., Levi, L., de Pena, R. G. & Norscini, R. (1961). Conductivité électrique de la glace dotée de divers électrolytes. *J. Chim. Phys.* **58**, 208–15. [159, 223, 225]

Itagaki, K. (1964). Self-diffusion in single crystals of ice. *J. Phys. Soc. Japan* **19**, 1081. [160]

Jaccard, C. (1959). Étude théorique et expérimentale des propriétés électriques de la glace. *Helv. Phys. Acta* **32**, 89–128. [118, 159, 212–15, 220, 234, 237]

Jaccard, C. (1963). Thermoelectric effects in ice crystals. *Phys. kondens. Materie* **2**, 143–51. [225, 230]

Jaccard, C. (1964). Thermodynamics of irreversible processes applied to ice. *Phys. kondens. Materie* **3**, 99–118. [212, 218, 230]

Jaccard, C. (1965). Mechanism of the electrical conductivity in ice. *Ann. New York Acad. Sci.* **125**, 390–400. [212]

Jaccard, C. (1966). Solute segregation at the curved surface of a growing crystal (steady state). *Phys. kondens. Materie* **4**, 349–54. [119]

Jaccard, C. (1967). Electrical conductivity of the surface layer of ice. In *Physics of Snow and Ice* (ed. H. Ôura), pp. 173–9. Japan: Institute of Low Temperature Science, Hokkaido Univ. [243]

Jaccard, C. & Levi, L. (1961). Ségrégation d'impuretés dans la glace. *Zeits. ang. Math. Phys.* **12**, 70–6. [119]

Jackson, K. A. (1958). Interface structure. In *Growth and Perfection of Crystals* (ed. R. H. Doremus, B. W. Roberts and D. Turnbull), pp. 319–24. New York: Wiley. [108]

Jackson, K. A. (1966). A review of the fundamental aspects of crystal growth. In *Crystal Growth* (ed. H. S. Peiser), pp. 17–24. Oxford: Pergamon. [108, 114]

Jakob, M. & Erk, S. (1928). Wärmedehnung des Eises zwischen 0 und −253°. *Z. ges. Kälte-Ind.* **35**, 125–30. [131]

Jakob, M. & Erk, S. (1929). Die Wärmeleitfähigkeit von Eis zwischen 0 und −125°. *Z. Tech. Phys.* **10**, 623–4. [143–4]

James, D. W. (1967). Solidification kinetics of ice determined by the thermal-wave technique. In *Crystal Growth* (ed. H. S. Peiser), pp. 767–73. Oxford: Pergamon. [114]

Jellinek, H. H. G. (1961). Liquidlike layers on ice. *J. Appl. Phys.* **32**, 1793. [127]

Jellinek, H. H. G. & Brill, R. (1956). Viscoelastic properties of ice. *J. Appl. Phys.* **27**, 1198–209. [187, 197]

Jhon, M. S., Grosh, J., Ree, T. & Eyring, H. (1966). Significant-structure theory applied to water and heavy water. *J. Chem. Phys.* **44**, 1465–72. [83]

Johnston, W. G. (1962). Yield points and delay times in single crystals. *J. Appl. Phys* **33**, 2716–30. [190, 195]

Jona, F. & Scherrer, P. (1952). Die elastischen Konstanten von Eis-Einkristallen. *Helv. Phys. Acta* **25**, 35–54. [118, 169–70]

Jones, S. J. (1967). Softening of ice crystals by dissolved fluoride ions. *Phys. Letters* **25**A, 366–7. [194]

Joshi, S. K. (1961). On the quasi-crystalline model of water. *J. Chem. Phys.* **35**, 1141–2. [76]

Kamb, B. (1961). The glide direction in ice. *J. Glaciol.* **3**, 1097–106. [195]

Kamb, B. (1964). Ice II: A proton-ordered form of ice. *Acta Cryst.* **17**, 1437–49. [62–3]

Kamb, B. (1965*a*). Structure of Ice VI. *Science, N.Y.* **150**, 205–9. [57, 68, 70]

Kamb, B. (1965*b*). Overlap interaction of water molecules. *J. Chem. Phys.* **43**, 3917–24. [47, 68, 71–2, 163]

Kamb, B. (1968). Ice polymorphism and the structure of liquid water. In *Structural Chemistry and Molecular Biology*, ed. A. Rich and N. Davidson. San Francisco: Freeman. [26, 59, 65, 67, 69, 72]

Kamb, B. & Datta, S. K. (1960). Crystal structures of the high-pressure forms of Ice: Ice III. *Nature, Lond.* **187**, 140–1. [64]

Kamb, B. & Davis, B. L. (1964). Ice VII, the densest form of ice. *Proc. Natn. Acad. Sci. U.S.A.* **52**, 1433–9. [71]

Kamb, B., Prakash, A. & Knobler, C. (1967). Structure of Ice V. *Acta Cryst.* **22**, 706–15. [66]

Kavanau, J. L. (1964). *Water and Solute-Water Interactions.* San Francisco: Holden-Day. [73, 97]

Kell, G. S. & Whalley, E. (1968). The equilibrium line between Ice I and III. *J. Chem. Phys.* **48**, 2359–61. [66]

Kevan, L. (1965). Cation interactions of trapped electrons in irradiated alkaline ice. *J. Am. Chem. Soc.* **87**, 1481–3. [164]

Kim, D.-Y. & Schmidt, V. H. (1967). Semiclassical theory of proton transport in ice. *Can. J. Phys.* **45**, 1507–16. [242]

Kingery, W. D. (ed.) (1963). *Ice and Snow.* Cambridge, Mass.: M.I.T. Press. [196]

Kittel, C. (1966). *Introduction to Solid State Physics.* New York: Wiley. [145–6]

Kneser, H. O., Magun, S. & Ziegler, G. (1955). Mechanischen Relaxation von einkristallinen Eis. *Naturwissenschaften* **42**, 437. [174, 182]

Knight, C. A. (1962). Curved growth of ice on surfaces. *J. Appl. Phys.* **33**, 1808–15. [114]

Knight, C. A. (1966). Grain boundary migration and other processes in the formation of ice sheets on water. *J. Appl. Phys.* **37**, 568–74. [111]

Kobayashi, T. (1958). On the habit of snow crystals artificially produced at low pressures. *J. Met. Soc. Japan*, ser. 2, **36**, 193–208 [121–2]

Kopp, M., Barnaal, D. E. & Lowe, I. J. (1965). Measurement by NMR of the diffusion rate of HF in ice. *J. Chem. Phys.* **43**, 2965–71. [161]

Korst, M. N., Savel'ev, V. A. & Sokolov, N. D. (1964). Second moment of the NMR signal and the structure of ice. *Soviet Phys. Solid St.* **6**, 965–6. [33]

Kramer, J. J. & Tiller, W. A. (1962). Determination of the atomic kinetics of the freezing process. *J. Chem. Phys.* **37**, 841–8. [116]

Kuhn, W. & Thürkauf, M. (1958). Isotopentrennung beim Gefrieren von Wasser und Diffusionskonstanten von D und ^{18}O im Eis. *Helv. chim. Acta* **41**, 938–71. [159–60]

Kume, K. & Hoshino, R. (1961). Proton magnetic resonance of ice single crystal. *J. Phys. Soc. Japan* **16**, 290–2. [33]

Kuroiwa, D. (1964). Internal friction of ice. *Contr. Inst. Low Temp. Sci., Hokkaido Univ.* A**18**, 1–62. [174, 183–4]

Kuroiwa, D. & Yamaji, K. (1959). Internal friction of polycrystalline and single-crystal ice. *Low Temp. Sci.* A**18**, 97–114. [173–4]

Lamb, J. & Turney, A. (1949). The dielectric properties of ice at 1·25 cm wavelength. *Proc. Phys. Soc.* B**62**, 272–3. [198]

Langham, E. J. & Mason, B. J. (1958). The heterogeneous and homogeneous nucleation of supercooled water. *Proc. Roy. Soc. Lond.* A**247**, 493–504. [86]

Latham, J. (1963). Charge transfer associated with temperature gradients in ice. *Nature, Lond.* **200**, 1087. [226]

Latham, J. (1964). Charge transfer associated with temperature gradients in ice crystals grown in a diffusion chamber. *Q. Jl Roy. Met. Soc.* **90**, 266–74. [227]

Latham, J. & Mason, B. J. (1961*a*). Electric charge transfer associated with temperature gradients in ice. *Proc. Roy. Soc. Lond.* A**260**, 523–36. [226, 228]

Latham, J. & Mason, B. J. (1961*b*). Generation of electric charge associated with the formation of soft hail in thunderclouds. *Proc. Roy. Soc. Lond.* A**260**, 537–49. [227]

LaPlaca, S. & Post, B. (1960). Thermal expansion of ice. *Acta Cryst.* **13**, 503–5. [26–7, 131]

Leadbetter, A. J. (1965). The thermodynamic and vibrational properties of H_2O ice and D_2O ice. *Proc. Roy. Soc. Lond.* A **287**, 403–25. [133, 141, 143]

Levi, L. & Arias, D. (1964). Continuous current conductivity of ice doped with various hydracids. *J. Chim. Phys.* **61**, 668–71. [225]

Levi, L. & Lubart, L. (1961). Electrical conductivity of ice doped with NH_4OH. *J. Chim. Phys.* **58**, 863–8. [159, 223]

Levi, L., Milman, O. & Suraski, E. (1963). Electrical conductivity and dissociation constants in ice doped with HF and NH_3 in different ratios. *Trans. Faraday Soc.* **59**, 2064–75. [150–1, 159, 223–4]

Lindenmeyer, C. S. & Chalmers, B. (1966*a*). Morphology of ice dendrites. *J. Chem. Phys.* **45**, 2804–6. [113–14]

Lindenmeyer, C. S. & Chalmers, B. (1966*b*). Growth rate of ice dendrites in aqueous solutions. *J. Chem. Phys.* **45**, 2807–8. [116, 118]

Lindenmeyer, C. S., Orrok, G. T., Jackson, K. A. & Chalmers, B. (1957). Rate of growth of ice crystals in supercooled water. *J. Chem. Phys.* **27**, 822. [114]

Lippincott, E. R., Finch, J. N. & Schroeder, R. (1959). Potential function model of hydrogen bond systems. In *Hydrogen Bonding* (ed. D. Hadzi), pp. 361–74. London: Pergamon Press. [43]

Lippincott, E. R. & Schroeder, R. (1955). One-dimensional model of the hydrogen bond. *J. Chem. Phys.* **23**, 1099–106. [43–5, 147]

London, F. (1937). The general theory of molecular forces. *Trans. Faraday Soc.* **33**, 8–26. [21]

Long, E. A. & Kemp, J. D. (1936). The entropy of deuterium oxide and the third law of thermodynamics. *J. Am. Chem. Soc.* **58**, 1829–34. [37, 134–5]

Longuet-Higgins, H. C. (1965). Intermolecular forces. *Disc. Faraday Soc.* **40**, 7–18. [22]

Lonsdale, K. (1958). The structure of ice. *Proc. Roy. Soc. Lond.* A **247**, 424–34. [26–7]

Lothe, J. & Pound, G. M. (1962). Reconsiderations of nucleation theory. *J. Chem. Phys.* **36**, 2080–5. [88]

McDonald, J. E. (1953). Homogeneous nucleation of supercooled water. *J. Met.* **10**, 416–33. [96]

MacDonald, J. R. (1955). Note on theories of time-varying space-charge polarization. *J. Chem. Phys.* **23**, 2308–9. [210]

McFarlan, R. L. (1936*a*). The structure of Ice II. *J. Chem. Phys.* **4**, 60–4. [64]

McFarlan, R. L. (1936*b*). The structure of Ice III. *J. Chem. Phys.* **4**, 243–9. [64]

McWeeny, R. & Ohno, K. A. (1960). A quantum-mechanical study of the water molecule. *Proc. Roy. Soc. Lond.* A **255**, 367–81. [5, 9, 46]

Macklin, W. C. & Ryan, B. F. (1965). The structure of ice grown in bulk supercooled water. *J. Atmos. Sci.* **22**, 452–9. [112–13, plate 2]

Macklin, W. C. & Ryan, B. F. (1966). Habits of ice grown in supercooled water and aqueous solutions. *Phil. Mag.* **14**, 847–60. [112–13]

Magono, C. & Lee, C. W. (1966). Meteorological classification of natural snow crystals. *J. Fac. Sci. Hokkaido Univ.* ser. VII, **2**, 321–35. (27 plates.) [119, plate 1]

Malenkov, G. G. (1961). Contribution to the problem of the structure of liquid water. *Soviet Phys. Dokl.* **6**, 292–3. [84]

Marchi, R. P. & Eyring, H. (1964). Application of significant structure theory to water. *J. Phys. Chem.* **68**, 221–8. [83]

Mason, B. J. (1953). The growth of ice crystals in a supercooled water cloud. *Q. J. Roy. Met. Soc.* **79**, 104–11. [128]

Mason, B. J., Bryant, G. W. & Van den Heuvel, A. P. (1963). The growth habits and surface structure of ice crystals. *Phil. Mag.* **8**, 505–26. [123–4]

Mason, B. J. & Van den Heuvel, A. P. (1959). The properties and behaviour of some artificial ice nuclei. *Proc. Phys. Soc.* **74**, 744–55. [102]

Mills, I. M. (1963). Force constant calculations for small molecules. In *Infra-red Spectroscopy and Molecular Structure* (ed. M. Davies), pp. 166–98. Amsterdam: Elsevier Pub. Co. [14]

Moccia, R. (1964). One-center basis set SCF MO's III. H_2O, H_2S and HCl. *J. Chem. Phys.* **40**, 2186–92. [3]
Moorthy, P. N. & Weiss, J. J. (1964). Formation of colour centres in irradiated alkaline ice. *Phil. Mag.* **10**, 659–74. [164]
Morgan, J. & Warren, B. E. (1938). X-ray analysis of the structure of water. *J. Chem. Phys.* **6**, 666–73. [74–5]
Muguruma, J. (1961). Spiral etch-pits of ice crystals. *Nature, Lond.* **190**, 37–8. [117]
Muguruma, J. & Higashi, A. (1963*a*). Observations of etch channels on the (0001) plane of ice crystal produced by nonbasal glide. *J. Phys. Soc. Japan* **18**, 1261–9. [196]
Muguruma, J. & Higashi, A. (1963*b*). Non-basal glide bands in ice crystals. *Nature, Lond.* **198**, 573. [196]
Muguruma, J., Mae, S. & Higashi, A. (1966). Void formation by non-basal glide in ice single crystals. *Phil. Mag.* **13**, 625–9. [196]
Nagle, J. F. (1966). Lattice statistics of hydrogen-bonded crystals. I. The residual entropy of ice. *J. Math. Phys.* **7**, 1484–91. [37]
Nakaya, U. (1954). *Snow Crystals: Natural and Artificial.* Cambridge, Mass.: Harvard University Press. [24, 119–20]
Némethy, G. & Scheraga, H. A. (1962). Structure of water and hydrophobic bonding in proteins. I. A model for the thermodynamic properties of liquid water. *J. Chem. Phys.* **36**, 3382–400. [81–2]
Nye, J. F. (1951). The flow of glaciers and ice-sheets as a problem in plasticity. *Proc. Roy. Soc. Lond.* A **207**, 554–72. [196]
Nye, J. F. (1957). *Physical Properties of Crystals.* Oxford: Clarendon Press. [165, 168]
Nye, J. F. (1967). Theory of regelation. *Phil. Mag.* **16**, 1249–66. [197]
Ockman, N. (1958). The infra-red and Raman spectra of ice. *Adv. Phys.* **7**, 199–220. [29, 136–7]
Onaka, R. & Takahashi, T. (1968). Vacuum u.v. absorption spectra of liquid water and ice. *J. Phys. Soc. Japan* **24**, 548–50. [212]
Onsager, L. & Dupuis, M. (1960). The electrical properties of ice. *Rc. Scu. int. Fis. 'Enrico Fermi'*, Corso x, pp. 294–315. [212, 218, 240]
Onsager, L. & Runnels, L. K. (1963). Mechanism for self-diffusion in ice. *Proc. Natn. Acad. Sci. U.S.A.* **50**, 208–10. [160]
Orentlicher, M. & Vogelhut, P. O. (1966). Structure and properties of liquid water. *J. Chem. Phys.* **45**, 4719–24. [80]
Oriani, R. A. & Sundquist, B. E. (1963). Emendations to nucleation theory and the homogeneous nucleation of water from the vapour. *J. Chem. Phys.* **38**, 2082–9. [88]
Oster, G. & Kirkwood, J. G. (1943). The influence of hindered molecular rotation on the dielectric constants of water, alcohols and other polar liquids. *J. Chem. Phys.* **11**, 175–8. [205]
Ôura, H. (ed.) (1967). *Physics of Snow and Ice* (Proceedings of Sapporo Conference, 1966), vols. 1 and 2. Japan: Institute of Low Temperature Science, Hokkaido University. [196]
Owston, P. G. (1958). The structure of Ice-I, as determined by X-ray and neutron diffraction analysis. *Adv. Phys.* **7**, 171–88. [26]
Patel, J. R. & Chaudhuri, A. R. (1963). Macroscopic plastic properties

of dislocation-free germanium and other semiconductor crystals. *J. Appl. Phys.* **34**, 2788–99. [194]

Pauling, L. (1935). The structure and entropy of ice and of other crystals with some randomness of atomic arrangement. *J. Am. Chem. Soc.* **57**, 2680–4. [30, 35]

Pauling, L. (1959). The structure of water. In *Hydrogen Bonding* (ed. D. Hadzi), pp. 1–5. London: Pergamon Press. [84]

Pauling, L. (1960). *The Nature of the Chemical Bond.* Ithaca, N.Y.: Cornell University Press. [1, 69]

Penny, A. H. A. (1948). A theoretical determination of the elastic constants of ice. *Proc. Camb. Phil. Soc.* **44**, 423–39. [168, 170]

Peterson, S. W. & Levy, H. A. (1957). A single-crystal neutron diffraction study of heavy ice. *Acta Cryst.* **10**, 70–6. [32–3]

Pimentel, G. C. & McClellan, A. L. (1960). *The Hydrogen Bond.* San Francisco: Freeman. [39]

Pistorius, C. W. F. T., Pistorius, M. C., Blakely, J. P. & Admiraal, L. J. (1963). Melting curve of Ice VII to 200 kbar. *J. Chem. Phys.* **38**, 600–2. [54, 70]

Pople, J. A. (1950). The molecular orbital theory of chemical valency. V. The structure of water and similar molecules. *Proc. Roy. Soc. Lond.* A **202**, 323–36. [5, 9]

Pople, J. A. (1951). Molecular association in liquids. II. A theory of the structure of water. *Proc. Roy. Soc. Lond.* A **205**, 163–78. [77–8]

Pople, J. A. (1953). The electronic structure and polarity of the water molecule. *J. Chem. Phys.* **21**, 2234–5. [9]

Pounder, E. R. (1965). *The Physics of Ice.* London: Pergamon Press. [119, 196]

Powell, R. W. (1958). Thermal conductivities and expansion coefficients of water and ice. *Adv. Phys.* **7**, 276–97. [131, 144]

Powles, J. G. (1952). A calculation of the static dielectric constant of ice. *J. Chem. Phys.* **20**, 1302–9. [202, 205]

Powles, J. G. (1955). Extension of Fröhlich's theory of the static dielectric constant to dielectrically anisotropic materials. *Trans. Faraday Soc.* **51**, 377–82. [206]

Pruppacher, H. R. (1963). Some relations between the supercooling and the structure of aqueous solutions. *J. Chem. Phys.* **39**, 1586–94. [97]

Pruppacher, H. R. (1967*a*). Growth modes of ice crystals in supercooled water and aqueous solutions. *J. Glaciol.* **6**, 651–62. [112–13, 118]

Pruppacher, H. R. (1967*b*). On the growth of ice in aqueous solutions contained in capillaries. *Z. Naturforsch.* **22*a***, 895–901. [114, 118]

Pruppacher, H. R. (1967*c*). Interpretation of experimentally determined growth rates of ice crystals in supercooled water. *J. Chem. Phys.* **47**, 1807–13. [114, 117]

Pruppacher, H. R., Steinberger, E. H. & Wang, T. L. (1968). On the electrical effects that accompany the spontaneous growth of ice in supercooled aqueous solutions. *J. Geophys. Res.* **73**, 571–84. [244]

Pryde, J. A. (1966). *The Liquid State.* London: Hutchinson. [73]

Pryde, J. A. & Jones, G. O. (1952). Properties of vitreous water. *Nature, Lond.* **170**, 685–8. [60]

Ratcliffe, E. H. (1962). The thermal conductivity of ice. New data on the temperature coefficient. *Phil. Mag.* **7**, 1197–1203. [143–4]

Read, W. T. & Shockley, W. (1950). Dislocation models of crystal grain boundaries. *Phys. Rev.* **78**, 275–89. [101]

Ready, D. W. & Kingery, W. D. (1964). Plastic deformation of single crystal ice. *Acta Metall.* **12**, 171–8. [188, 190]

Reiss, H. & Katz, J. L. (1967). Resolution of the translation-rotation paradox in the theory of irreversible condensation. *J. Chem. Phys.* **46**, 2496–9. [88]

Rowlinson, J. S. (1951). The lattice energy of ice and the second virial coefficient of water vapour. *Trans. Faraday Soc.* **47**, 120–9. [10, 42, 45]

Samoilov, O. Ya. (1965). *Structure of Aqueous Electrolyte Solutions.* New York: Consultants Bureau. [97]

Schiller, P. (1958). Die mechanische Relaxation in reinen Eiseinkristallen. *Z. Phys.* **153**, 1–15. [174, 179]

Schmitt, R. G. & Brehm, R. K. (1966). Double beam spectrophotometry in the far ultraviolet. *Applied Optics* **5**, 1111–16. [15–16]

Schoeck, G. (1956). Moving dislocations and solute atoms. *Phys. Rev.* **102**, 1458–9. [194]

Schulz, H. (1961). Die mechanische Relaxation in Eis-HF-Mischkristallen verschiedener HF-Konzentration. *Naturwissenschaften.* **48**, 691. [183]

Shaw, D. & Mason, B. J. (1955). The growth of ice crystals from the vapour. *Phil. Mag.* **46**, 249–62. [128]

Shimaoka, K. (1960). Electron diffraction study of ice. *J. Phys. Soc. Japan* **15**, 106–19. [32, 58]

Siegel, S., Flournoy, J. M. & Baum, L. H. (1961). Irradiation yields of radicals in gamma-irradiated ice at 4·2° and 77° K. *J. Chem. Phys.* **34**, 1782–8. [164]

Singwi, K. S. & Sjölander, A. (1960). Diffusive motions in water and cold neutron scattering. *Phys. Rev.* **119**, 863–71. [76]

Slater, J. C. (1941). Theory of the transition in KH_2PO_4. *J. Chem. Phys.* **9**, 16–33. [245]

Slater, J. C. (1963). *Quantum Theory of Molecules and Solids.* Vol. 1. *Electronic Structure of Molecules.* New York: McGraw-Hill. [1]

Smyth, C. P. & Hitchcock, C. S. (1932). Dipole rotation in crystalline solids. *J. Am. Chem. Soc.* **54**, 4631–47. [209]

Somorjai, R. L. & Hornig, D. F. (1962). Double-minimum potentials in hydrogen-bonded solids. *J. Chem. Phys.* **36**, 1980–7. [239]

Sparnaay, M. J. (1966). Water structure and interionic interaction. *J. Colloid Interface Sci.* **22**, 23–31. [80]

Steinemann, A. (1957). Dielektrische Eigenschaften von Eiskristallen. *Helv. Phys. Acta* **30**, 581–610. [159, 161, 209–10, 219–21]

Steinemann, A. & Gränicher, H. (1957). Dielektrische Eigenschaften von Eiskristallen. *Helv. Phys. Acta* **30**, 553–80. [222]

Steinemann, S. (1953). Polar crystal form and piezo electricity of ice. *Experientia* **9**, 135–6. [30]

Steinemann, S. (1954). Results of preliminary experiments on the plasticity of ice crystals. *J. Glaciol.* **2**, 404–12. [187]

Sussman, J. A. (1964). Phonon induced tunnelling of ions in solids. *Phys. kondens. Materie* **2**, 146–60. [240]

Takahashi, T. (1966). Thermoelectric effect in ice. *J. Atmos. Sci.* **23**, 74–7. [227–8]

Taylor, M. J. & Whalley, E. (1964). Raman spectra of Ices, I_h, I_c, II, III and V. *J. Chem. Phys.* **40**, 1660–4. [61, 136–7]

Teichmann, I. & Schmidt, G. (1965). Investigations of the piezoelectric effect of ice. *Phys. Stat. Solidi* **8**, K145–7. [30]

Thomas, M. R. & Scheraga, H. A. (1965). A near-infrared study of hydrogen bonding in water and deuterium oxide. *J. Phys. Chem.* **69**, 3722–6. [82]

Turnbull, D. (1956). Phase changes. In *Solid State Physics* (ed. F. Seitz and D. Turnbull), vol. III, pp. 225–306. New York: Academic Press. [88]

Turnbull, D. & Fisher, J. C. (1949). Rate of nucleation in condensed systems. *J. Chem. Phys.* **17**, 71–3, 429. [88, 94]

Turnbull, D. & Vonnegut, B. (1952). Nucleation catalysis. *Ind. Engng Chem. (Industr.)* **44**, 1292–8. [101]

Ubbelohde, A. R. (1965). *Melting and Crystal Structure*, ch. 14. Oxford: Clarendon Press. [85]

Vand, V. & Senior, W. A. (1965). Structure and partition function of liquid water. I, II and III. *J. Chem. Phys.* **43**, 1869–84. [82]

Van der Merwe, J. H. (1950). On the stresses and energies associated with inter-crystalline boundaries. *Proc. Phys. Soc.* A **63**, 616–37. [101]

Van der Merwe, J. H. (1963). Crystal interfaces. I. Semi-infinite crystals; II. Finite overgrowths. *J. Appl. Phys.* **34**, 117–27. [101]

Volmer, M. (1939). *Kinetik der Phasenbilding*. Dresden and Leipzig: Steinkopff. [87]

Volmer, M. & Weber, A. (1926). Keimbildung in übersättigen Gebilden. *Z. Phys. Chem.* **119**, 277–301. [87]

Wall, T. T. & Hornig, D. F. (1965). Raman intensities of HDO and structure in liquid water. *J. Chem. Phys.* **43**, 2079–87. [79, 82]

Walrafen, G. E. (1966). Raman spectral studies of the effects of temperature on water and electrolyte solutions. *J. Chem. Phys.* **44**, 1546–58. [82]

Walrafen, G. E. (1968). Raman spectral studies of HDO in H_2O. *J. Chem. Phys.* **48**, 244–51. [80, 82]

Walz, E. & Magun, S. (1959). Die mechanische Relaxation in Eis-NH_4F-Mischkristallen. *Z. Phys.* **157**, 266–74. [183]

Watanabe, K. & Zelikoff, M. (1953). Absorption coefficients of water vapour in the vacuum ultraviolet. *J. Opt. Soc. Am.* **43**, 753–5. [15]

Webb, W. W. & Hayes, C. E. (1967). Dislocations and plastic deformation of ice. *Phil. Mag.* **16**, 909–25. [116, 192, plate 3]

Weertman, J. (1957). Steady-state creep of crystals. *J. Appl. Phys.* **28**, 1185–9. [192]

Weir, C., Block, S. & Piermarini, G. (1965). Single-crystal X-ray diffraction at high pressures. *J. Res. Nat. Bur. Stand.* **69** C, 275–81. [68, 71]

Weissmann, M. & Cohan, N. V. (1965*a*). Molecular orbital study of the hydrogen bond in ice. *J. Chem. Phys.* **43**, 119–23. [39, 40, 47, 148]

Weissmann, M. & Cohan, N. V. (1965*b*). Molecular orbital study of ionic defects in ice. *J. Chem. Phys.* **43**, 124–6. [152–3, 238–9]

Weyl, W. A. (1951). Surface structure of water and some of its physical and chemical manifestations. *J. Colloid Sci.* **6**, 389–405. [126]

Whalley, E. (1957). The difference in the intermolecular forces of H_2O and D_2O. *Trans. Faraday Soc.* **53**, 1578–85. [41]

Whalley, E. & Bertie, J. E. (1967). Optical spectra of orientationally disordered crystals. I. Theory for translational lattice vibrations. *J. Chem. Phys.* **46**, 1264–70. [138]

Whalley, E. & Davidson, D. W. (1965). Entropy changes at the phase transitions in ice. *J. Chem. Phys.* **43**, 2148–9. [66]

Whalley, E., Davidson, D. W. & Heath, J. B. R. (1966). Dielectric properties of Ice VII. Ice VIII: A new phase of ice. *J. Chem. Phys.* **45**, 3976–82. [50, 54, 56, 70]

Whalley, E., Heath, J. B. R. & Davidson, D. W. (1968). Ice IX: An antiferroelectric phase related to Ice III. *J. Chem. Phys.* **48**, 2362–70. [50, 56, 61, 66]

Whalley, E. & Labbé, H. J. (1969). Optical spectra of orientationally disordered crystals. III. Infrared spectra of the sound waves. *J. Chem. Phys.* (in the Press). [140]

Wilson, G. J., Chan, R. K., Davidson, D. W. & Whalley, E. (1965). Dielectric properties of Ices II, III, V and VI. *J. Chem. Phys.* **43**, 2384–91. [56, 61, 64, 70]

Woerner, S. & Magun, S. (1959). Mechanische Dämpfung von D_2O-Einkristallen. *Naturwissenschaften* **46**, 509–10. [183]

Wollan, E. O., Davidson, W. L. & Shull, C. G. (1949). Neutron diffraction study of the structure of ice. *Phys. Rev.* **75**, 1348–52. [32]

Workman, E. J. & Reynolds, S. E. (1950). Electrical phenomena occurring during the freezing of dilute aqueous solution and their possible relationship to thunderstorm electricity. *Phys. Rev.* **78**, 254–9. [244]

Wulff, G. (1901). Zur Frage der Geschwindigkeit des Wachtums und der Auflösung der Krystallflächen. *Z. Krystallogr.* **34**, 449–530. [122]

Yang, L. C. & Good, W. B. (1966). Crystallization rate of supercooled water in cylindrical tubes. *J. Geophys. Res.* **71**, 2465–9. [114]

Zajac, A. (1958). Thermal motion in ice and heavy ice. *J. Chem. Phys.* **29**, 1324–7. [143]

Zarembovitch, A. & Kahane, A. (1964). Détermination des vitesses de propagation d'ondes ultrasonores longitudinales dans la glace. *C. r. hebd. Séanc. Acad. Sci., Paris* **258**, 2529–32. [169–70, 172]

Zaromb, S. & Brill, R. (1956). Solid solutions of ice and NH_4F and their dielectric properties. *J. Chem. Phys.* **24**, 895–902. [224]

Zettlemoyer, A. C., Tcheurekdjian, N. & Hosler, C. L. (1963). Ice nucleation by hydrophobic substances. *Z. ang. Math. Phys.* **14**, 496–502. [100]

Ziman, J. M. (1960). *Electrons and Phonons*. Oxford University Press. [138, 145–6, 230]

Ziman, J. M. (1964). *Principles of the Theory of Solids*. Cambridge University Press.) [38]

SUBJECT INDEX